Nontechnical Attribute Training and Development for United States Air Force Special Warfare

SEAN ROBSON, TRACY C. KRUEGER, MARIA C. LYTELL, SUSAN G. STRAUS, LEWIS SCHNEIDER, ANTHONY ATLER, GRACE FALGOUST, MATT STRAWN, MARGARET A. MAGLIONE, ZHAN OKUDA-LIM, SAM WALLACE, MIRIAM MATTHEWS

Prepared for the Department of the Air Force
Approved for public release; distribution is unlimited.

RAND PROJECT AIR FORCE

For more information on this publication, visit **www.rand.org/t/RRA1747-1**.

Published by the RAND Corporation, Santa Monica, Calif.

Library of Congress Cataloging-in-Publication Data is available for this publication.

ISBN: 978-1-9774-1405-2

Cover: Carly Kavish/U.S. Air Force.

About This Report

A primary goal of this project is to offer a roadmap for developing Air Force Special Warfare (AFSPECWAR) attributes (which are nontechnical, such as Stress Tolerance and Teamwork) among its operators with a focus on what should be emphasized during initial skills training and other early career training courses. This line of effort supports other talent management initiatives such as assessment and selection (A&S) described in our work with AFSPECWAR and Air Force Special Operations Command (AFSOC) in fiscal years (FYs) 2020 and 2021.

AFSPECWAR is the ground combat force for the Department of the Air Force (DAF) and specializes in the application of airpower. AFSPECWAR consists of seven career fields: Combat Control (CCT), Pararescue (PJ), Tactical Air Control Party (TACP), Special Reconnaissance (SR), Combat Rescue Officer (CRO), Special Tactics Officer (STO), and TACP Officer (TACP-O). Among the many changes to talent management practices, this community implemented a new A&S course in January 2019. In FY 2020 and FY 2021, we focused on defining and evaluating AFSPECWAR attributes. This report focuses on the developing those same nontechnical attributes. Specifically, we review best practices for developing attributes, evaluate AFSPECWAR efforts to develop those attributes, and design a roadmap describing how attribute development can be incorporated into the training pipeline.

The research reported here was commissioned by Second Air Force (2AF) and conducted within the Workforce, Development, and Health Program of RAND Project AIR FORCE as part of a FY 2022 project, "Transforming Air Force Special Warfare Human Capital."

RAND Project AIR FORCE

RAND Project AIR FORCE (PAF), a division of RAND, is the DAF's federally funded research and development center for studies and analyses. PAF provides the DAF with independent analyses of policy alternatives affecting the development, employment, combat readiness, and support of current and future air, space, and cyber forces. Research is conducted in four programs: Strategy and Doctrine; Force Modernization and Employment; Resource Management; and Workforce, Development, and Health. The research reported here was prepared under contract FA7014-22-D-0001.

Additional information about PAF is available on our website:
www.rand.org/paf/

This report documents work originally shared with the DAF on October 3, 2022. The draft report, dated September 2022, was reviewed by formal peer reviewers and DAF subject-matter experts.

Acknowledgments

Many people made this project possible. We begin by thanking our original project sponsor, Maj Gen Andrea Tullos (Second Air Force, Commander), and her successor, Maj Gen Michele Edmondson, as well as Colonel (Col) Mason Dula (Special Warfare Training Wing [SWTW], Commander) who initiated this work. Col Dula and his entire leadership team strongly supported this project by being highly responsive, providing thoughtful feedback, and facilitating important connections. Specifically, we are grateful to Col Matthew Berry (SWTW, Vice Commander), Chief Master Sargent (CMSgt) Todd Popovic (SWTW, Commanders Chief), Col Robert Taylor (Special Warfare Training Group [SWTG], Commander), and CMSgt Nathan Cox (SWTG, Senior Enlisted Leader).

In addition to leadership, we thank our primary points of contact: Maj Emily Grieser, Lt Col Danielle Galt, and Dr. Karal Garcia. We also thank operational psychologists—Maj Jeffrey Geddes, Dr. Shawn Knadler, and Dr. Annamari Maaranen—for their insight and assistance.

As will be described in the report, this work required support from other Major Commands such as Air Force Special Operational Command (AFSOC) and Air Combat Command (ACC). We extend our gratitude to these leaders: Col Matthew Allen (24th Special Operations Wing [24 SOW], former Commander); Col Kenneth Daniels (24 SOW, Commander); Col Danielle Willis (93rd Air Ground Operations Wing [AGOW], Commander); and Col Peter White (563rd Rescue Group [RQG], Commander).

We could not have completed our focus groups without local points of contact providing subject-matter expertise as well as facilitating and coordinating various logistical needs. In particular, we thank Lt Col Matthew Arnold, Technical Sergeant(TSgt) Alberto Fernandezbravo, Major (Maj) John Wylie, Maj Amileah Davis, John Thomas, Lieutenant Colonel (Lt Col) Patrick Parrish, Senior Master Sergeant (SMSgt) Francisco Corona, Maj Daniel Hill, Maj Evan Cumming, TSgt Kyle Murray, Maj Zachary Ayscue, Captain (Capt) Matthew Peabody, Capt Christopher Cadieux, TSgt Steven Luckert, Lt Col Matthew McMurtry, CMSgt Jason Hammel, Maj Joshua Armstrong, Maj Mark Taylor, MSgt Seth Babin, and Capt Carl Schleich. There were nearly 450 participants representing a range of perspectives from the operational units, training pipeline, and individuals who self-initiated eliminated (SIE). This level of engagement is indicative of the overall commitment of AFSPECWAR. Thank you.

Several RAND colleagues also contributed to this project. Stephanie Williamson, Anthony Lawrence, Judith Mele, Paul Emslie, and John Crown assisted with acquiring, cleaning, de-identifying, and merging several data sources. Nelson Lim and David Schulker provided constructive feedback on interim project briefings and prior versions of this report. Finally, we would like to thank our reviewers, Steve Oliver and Stephen Zaccaro, for their constructive review and feedback.

Summary

Issue

Military training is a fundamental talent management activity that prepares warfighters to execute occupational and mission critical tasks. Initial skills training for Air Force Special Warfare (AFSPECWAR) is designed to efficiently teach unskilled airmen the technical knowledge and skills required to perform a particular occupational specialty. However, there is an emerging need to also develop the nontechnical attributes (e.g., Teamwork, Stress Tolerance) required to effectively learn and perform jobs in an increasingly dynamic and uncertain future.

Approach

We used a mixed-methods approach that included (1) a review of relevant documents and content analysis of training objectives and events described in course lesson plans; (2) discussions with course leadership, instructors, and students on how much emphasis training courses placed on developing AFSPECWAR attributes; (3) a needs analysis to identify which attributes may need further attention and development; (4) a targeted literature review to describe science-backed interventions; and (5) integration of research and assessment of Special Warfare Training Wing (SWTW) practices to develop a roadmap for nontechnical attribute development.

Key Findings

- AFSPECWAR is developing the eight AFSPECWAR nontechnical attributes. However, this development is not deliberate and occurs primarily through exposure to technical training instruction.
- The majority of students in AFSPECWAR training are performing well, but gaps were identified in four attributes:
  - Drive
  - Teamwork
  - Trainability
  - Stress Tolerance.
- Effective interventions should focus both on the individual (e.g., goal-setting) and the training environment (e.g., establishing and maintaining a strong culture by encouraging prompt and constructive feedback and appropriate reinforcements for performance improvements).
- Research suggests that attribute development should be increasingly job-specific as students advance through the training pipeline.

Recommendations

Drawing on our analyses, we developed a roadmap with seven recommendations to make AFSPECWAR's efforts to develop its nontechnical attributes more deliberate. Figure S.1 presents this general implementation roadmap that can be adopted by any functional community. Specifically, for AFSPECWAR, we emphasize the following steps to address the attribute gaps (Drive, Teamwork, Trainability, Stress Tolerance): (1) teach relevant frameworks as applicable to current course progression in the pipeline, (2) allow students to practice and apply the frameworks, and then (3) use credible sources to provide feedback on what is working and areas for additional development. More specific strategies to address attribute gaps should be considered based on a thorough evaluation of current practices and student needs.

Figure S.1. Recommendation Roadmap for Nontechnical Attribute Development

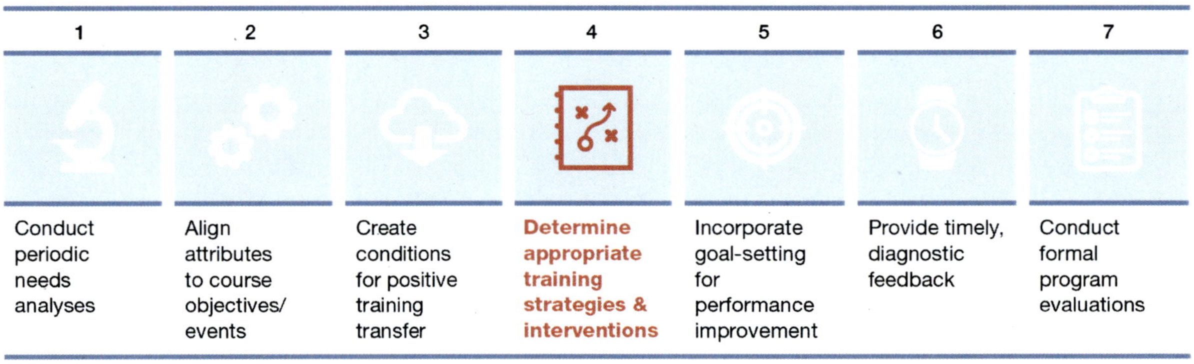

Contents

Figures and Tables

Figures

Tables

Chapter 1

Introduction

Military training is a fundamental talent management activity that prepares warfighters to execute occupational and mission critical tasks. Initial skills training in the United States Air Force (USAF) is designed to efficiently teach unskilled airmen the technical knowledge and skills required to perform a particular occupational specialty. However, this focus on technical training may miss opportunities to develop nontechnical attributes (e.g., Teamwork, Stress Tolerance) required for effective job performance. Recognizing the importance of these nontechnical attributes to prepare for future missions, Air Force Special Warfare (AFSPECWAR)[1] has supported efforts to improve its talent management practices.

As part of its foundational mission, the Special Warfare Training Wing (SWTW) is conducting a Pipeline Development Research Campaign to improve the quality and efficiency of the AFSPECWAR production. This campaign seeks to improve AFSPECWAR production by holistically evaluating all aspects, from recruiting to fielding of AFSPECWAR operators. In fiscal year (FY) 2020, RAND collaborated with AFSPECWAR to establish a common set of nontechnical attributes across the AFSPECWAR enterprise (e.g., Teamwork, Stress Tolerance).[2] In FY 2021, RAND evaluated the assessment and selection (A&S) course (e.g., effectiveness and utility of current assessments) as well as the strategic utilization of the activities preceding and following the A&S course.[3] This project continued the Pipeline Development Research Campaign by establishing a roadmap for developing AFSPECWAR nontechnical attributes.

Project Objectives and Approach

The focus of this report is on the eight AFSPECWAR attributes. Adapting an organizing framework, we aligned the AFSPECWAR attributes into four domains: (1) cognitive, (2) physical, (3) intrapersonal, and (4) interpersonal. (See Figure 1.1.)

Entry to AFSPECWAR consists of multiple phases that vary by occupational specialty (see Figure 1.2). Although nontechnical attribute development could occur during the recruitment and

---

[1] AFSPECWAR is comprised of four enlisted specialties and three officer career specialties. Enlisted specialties are Combat Control (CCT), Pararescue (PJ), Special Reconnaissance (SR), and Tactical Air Control Party (TACP). Officer career specialties are Combat Rescue Officer (CRO), Special Tactics Officer (STO), and Tactical Air Control Party-Officer (TACP-O).

[2] Tracy C. Krueger, Sean Robson, Joshua Snoke, Matthew Walsh, Matthew Strawn, Anthony Atler, Jason H. Campbell, Isabel Leamon, Ryan Haberman, and Barbara A. Bicksler, *Assessment and Selection for U.S. Air Force Special Warfare*: Vol. 1, *Defining Attributes and Designing Rater Training*, Santa Monica, Calif.: RAND, RR-A549-1, 2022.

[3] Tracy C. Krueger, Sean Robson, Matthew Walsh, Anthony Atler, Beverly Weidmer, Matthew Strawn, Lewis Schneider, Lisa Pelled Colabella, Yuliya Shokh, and Tiffany Berglund, *Assessment and Selection for U.S. Air Force Special Warfare*: Vol. 2, *Evaluating Objectives and Considering Improvements*, Santa Monica, Calif.: RAND, RR-A1199-1, 2024.

Figure 1.1. Organizing Framework for the Eight AFSPECWAR Attributes

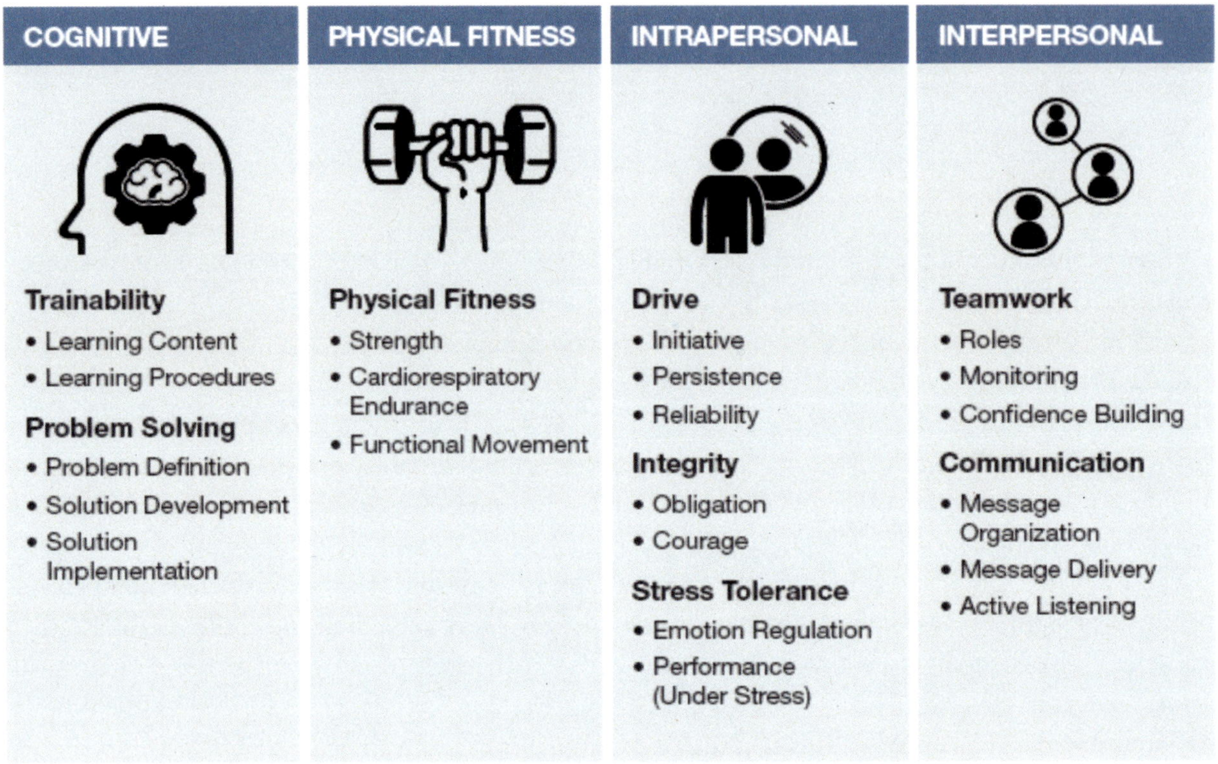

Figure 1.2. Overview of AFSPECWAR Pipeline

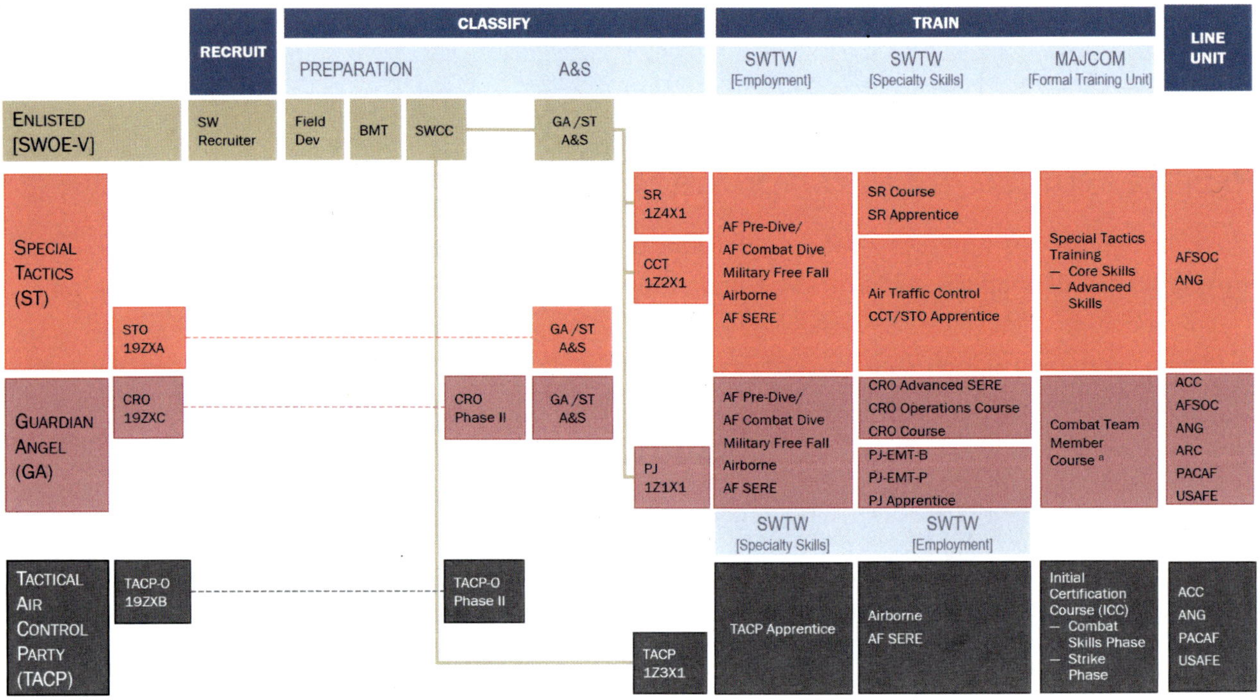

NOTE: This overview is current as of June 2022. ACC = Air Combat Command; PACAF = Pacific Air Forces; USAFE = Europe–Air Forces Africa; ANG = Air National Guard; ARC = Air Reserve Component.

classification stages, we focused on the apprentice courses managed by SWTW and select Major Command (MAJCOM) formal training units (FTU), including Special Tactics Training, the Tactical Air Control Party (TACP) Initial Certification Course, and Combat Team Member Course. Because formal and informal development that may occur once on the job were outside the scope of this project, we do not focus on certain streams of research that would be relevant as operators mature in the career field (e.g., leader development, developmental job assignments).

Project Questions and Our Approach

Continuing to address key issues in the research campaign, Second Air Force (2AF) and the SWTW asked RAND Project AIR FORCE (PAF) to address the following questions:

- What are the USAF and SWTW currently doing to develop nontechnical attributes? (Chapter 2)
- What are SWTW's training needs for developing nontechnical attributes? (Chapter 3)
- What best practices exist to develop cognitive attributes: Trainability and Problem Solving? (Chapter 4)
- What best practices exist to develop Physical Fitness? (Chapter 5)
- What best practices exist to develop intrapersonal attributes: Drive, Integrity, and Stress Tolerance? (Chapter 6)
- What best practices exist to develop interpersonal attributes: Teamwork and Communication? (Chapter 7)
- What should SWTW do to better support the development of nontechnical attributes during training? (Chapter 8)

To address these questions, we used a mixed-methods approach. We first established a baseline for how SWTW is developing attributes by reviewing relevant documents and conducting a content analysis of training objectives and events described in course lesson plans. We supplemented these analyses through discussions with course leadership, instructors, and students. These discussions focused on how much emphasis training courses placed on developing AFSPECWAR attributes. Next, we conducted a needs analysis to identify which attributes may need further attention and development. To help position SWTW to address these gaps and further develop all AFSPECWAR attributes, we conducted a targeted literature review to describe science-backed interventions. Finally, we integrated this research with our assessment of SWTW practices to develop a roadmap for nontechnical attribute development that could be generalized to the broader DAF.

Scope and Limitations

Based on discussions with SWTW leadership, we prioritized the courses for inclusion in the document review and for evaluations and discussions with instructors and students. We focus on training prior to an operator's first assignment and do not cover every course; therefore, some courses may provide development or additional opportunities to develop attributes. For example, the Survival,

Evasion, Resistance, and Escape (SERE) course—which was not included in our analysis—likely provides additional opportunities for students to practice techniques to enhance Stress Tolerance.

To review best practices for developing nontechnical attributes, we focused on systematic and narrative reviews that combine studies from different populations. Although we include a discussion of relevant military studies, when possible, some research findings may not generalize well to AFSPECWAR or to the military training environment. Finally, our needs analysis is cross-sectional and does not include longitudinal data following students over time. Therefore, some identified "deficiencies" among students may not require training interventions but could be resolved over time as students progress through the pipeline, allowing them the time needed to develop and mature. We also cannot account for changes that may have occurred during the project to address performance concerns. SWTW conducts regular reviews to address training content, delivery, and execution. Therefore, we strongly encourage engaging with current training instructors and course designers to verify whether these needs continue to exist or if other needs have emerged.

Organization of This Report

The remainder of this report consists of seven additional chapters followed by three appendixes. The next chapter (Chapter 2) describes current USAF and SWTW efforts to develop nontechnical attributes. Chapter 3 summarizes results from a needs analysis that identifies performance concerns among AFSPECWAR students. The next three chapters review specific lines of research dedicated to each attribute. Chapter 4 covers research on cognitive attributes including Trainability and Problem Solving. Chapter 5 reviews research on Physical Fitness. Chapter 6 addresses the three intrapersonal attributes: Drive, Integrity, and Stress Tolerance. Chapter 7 covers interpersonal attributes, which include Teamwork and Communication. We provide conclusions and recommendations in Chapter 8. We also include the following appendixes with additional information on project approach and analyses:

- Appendix A provides an overview of crosscutting features of training design and lays the foundation for many concepts that are reviewed in more detail throughout this report.
- Appendix B describes the methodology and supplementary analyses of the SWTW document review.
- Appendix C presents the methodology and supplementary analyses of the attribute literature review.

Chapter 2

Current Air Force Practices to Develop Nontechnical Attributes

This chapter describes current Department of the Air Force (DAF) and SWTW practices to develop nontechnical attributes. We first provide an overview of broader DAF efforts followed by a more detailed analysis of SWTW efforts, including results from a survey capturing instructor and student perspectives on how much AFSPECWAR attributes are emphasized in training courses. Next, we present the results of a content analysis that aligned training objectives and events to AFSPECWAR attributes. This analysis shows where training may directly or indirectly develop attributes. We conclude by comparing findings across sources to determine which attributes are currently emphasized in AFSPECWAR training.

DAF Efforts to Develop Nontechnical Attributes

Foundational Competencies

Foundational competencies are the knowledge, skills, abilities, and other characteristics (KSAOs) determined to be relevant across a wide variety of job types within the USAF. Updated and validated in 2019, there are 22 competencies deemed important for all airmen from E-1 to O-10 and WG-01 to Senior Executive Service (SES). Airmen are expected to develop these competencies throughout their careers, "including deliberate development on the Foundational Competencies during Basic Military Training, Professional Military Education, and civilian developmental education programs."[4] The competencies are as follows: accountability, analytical thinking, change management, communication, creative thinking, decisionmaking, develops people, flexibility, fostering innovation, influence, information seeking, initiative, leadership, perseverance, precision, resilience, resource management, results focused, self-control, service mindset, strategic thinking, and teamwork.

To support the development of these foundational competencies, efforts were made to create assessments ("kneeboard" rubrics and 66-item Likert-type measures) to evaluate the level of each competency an airman currently possesses. "Kneeboard" rubrics consist of nine- to ten-item assessments of each competency that distinguish behaviors of increasing levels of competency proficiency (Basic, Intermediate, Expert, and Advanced) and were developed for classroom use by trainers or supervisors.[5] The 66-item Likert-type measures are briefer than the "kneeboard" rubrics

---

[4] HQ AETC A3/6, *Development of Air Force Foundational Competency Assessments Technical Report No. 2*, p. 2.

[5] For example, the "accountability" competency consists of nine items broken into proficiency levels: "Leads by example" (expert); "models professionalism and excellence in every endeavor" (expert); "takes personal responsibility for unit performance" (expert); "does the right thing even when it is unpopular or difficult" (advanced); "admits shortcomings and mistakes" (advanced); "follows

and were developed for use in MyVector, a platform that supports multisource feedback among other tools for career development and will be discussed further in the next section.[6]

This multisource system is not currently used across AFSPECWAR. Depending on AFSPECWAR's specific goals for developing airmen, consideration of foundational competency assessments and the existing systems used to assess airmen could be used as a starting point for evaluating and providing feedback. However, to ensure relevance to AFSPECWAR's specific training and operational environments, additional steps would need to be taken to evaluate, tailor, or expand on "kneeboard" rubrics (or items). If implemented, collecting feedback at multiple points in time (e.g., annually) could help AFSPECWAR to track any improvements.

MyVector

MyVector is a career development application that allows for the tracking of competency development. Right now, airmen can take a self-assessment to gauge their current levels on the USAF foundational competencies. This self-assessment asks airmen to rate their level of agreement with statements that represent the different competencies (e.g., "I consistently and effectively do the right thing even when it is unpopular or difficult"). After taking the assessment, airmen receive feedback on competency strengths and areas of improvement. They also have the option to select a 360-feedback tool to receive confidential feedback from peers, subordinates, or higher-ranking members on their competency development. New competencies can be added to MyVector, and items used to assess those competencies can also be added. Questionnaires can be assembled using these items to assess new competencies deemed necessary for proper career development.

Relevant DAF Policies, Manuals, and Instructions

Basic Military and Technical Training

Air Education and Training Command Instruction 36-2651 (AETCI 36-2651) outlines "the roles, responsibilities, and programs required for executing and supporting formal enlisted entry-level accessions (basic military); officer, enlisted, and civilian non-flying occupational (technical) training programs; operational medicine training programs; and military standards training." The publication "applies to the Total Force–Regular Air Force, Air Force Reserve, Air National Guard, and Department of the Air Force (DAF) civilians permanently or temporarily assigned to the Air Education and Training Command (AETC)."[7]

Responsibilities are distributed by directorates and units. For example, the Directorate of Operations and Communications "develops and executes competency modeling processes and partners with functional authorities, functional managers, career field managers, and education and training specialists to guide, measure, implement, assess, and oversee all aspects of competency modeling for the

---

through on promises and commitments" (intermediate); "embodies the Air Force Core Values of Integrity, Service Before Self, and Excellence" (intermediate); "adheres to Air Force standards" (basic); and "looks after fellow Airmen and their families" (basic).

[6] MyVector items were chosen based on their ability to capture the entire competency while remaining simple and easy to understand. For example, the "initiative" competency consists of the items "demonstrates an inspired work ethic," "exceeds job description (e.g., takes on extra tasks)," and "puts in extra effort to complete work when not required."

[7] U.S. Air Force, Basic Military and Technical Training, Air Education and Training Command Instruction 36-2651, April 12, 2021.

Air Force," among many other roles and responsibilities.[8] There are directorates in charge of training requirements, including personnel resources and instructors, and other directors monitoring policy and resourcing decisions, logistics support for training, training plans, training analysis and innovation, and training safety. This document provides task statements for each unit but does not go into detail about how these tasks are executed. As a further example, the Air Force Career Development Academy

- [m]anages the USAF career development course program
- [d]evelops and delivers interactive career development courses in support of the enlisted upgrade training program and the Weighted Airman Promotion System
- [e]stablishes and maintains publications, forms, and/or guides to identify roles, responsibilities, and processes for the creation, delivery, revision, and evaluation of career development courses and specialized courses in accordance with AFI 36-2670, Total Force Development
- [p]rovides a dashboard reporting real-time status of career development courses in development and completion statistics for implemented courseware.[9]

Information for Designers of Instructional Systems

The Air Force Handbook 36-2235 (Volume 9) "provides specific information and guidance for using the instructional system development (ISD) process to develop technical training."[10] ISD is the process used to design, develop, implement, and manage technical training in the USAF. This process is comprised of several functions that serve to provide effective training and functions necessary for process and training quality improvement.

The ISD model consists of multiple functions and processes:

- System functions
  - Management: the function of directing or controlling instructional system development and operations
  - Support: the function of maintaining all parts of the system
  - Administration: the function of day-to-day processing and recordkeeping
  - Delivery: the function of bringing instruction to students
  - Evaluation: the function of gathering feedback data through formative, summative, and operational evaluations to assess system and student performance.[11]
- Instructional development process
  - Analyze and determine what instruction is needed
  - Design instruction to meet the need
  - Develop instructional materials to support system requirements
  - Implement the instructional system

[8] U.S. Air Force, 2021.

[9] U.S. Air Force, 2021, p. 9.

[10] U.S. Air Force, Information for Designers of Instructional Systems, Air Force Handbook 36-2235, November 1, 2002, p. 4.

[11] U.S. Air Force, 2002, p. 9.

- Evaluate, which is a central function that takes place at every phase.[12]

During the analysis process, occupational job and task analysis is conducted to understand the KSAOs needed to successfully perform the job. Learning outcomes can then be designed based on the desired tasks, KSAOs, and desired level of learning. Once this information is gathered, the training methodology that is most appropriate can be determined, and a training plan can be developed. Each of these processes is outlined in detail within this RAND report.

How to design training activities is also broken down in this document, and guidelines for how and when to use specific methodologies is provided. A variety of presentation, instruction, and knowledge application methods are explained, including different methods of questioning, discussion, simulation, and case studies. Understanding if the training is working as planned is necessary as well. The steps to develop well-constructed tests are outlined, and different types of tests are explained, primarily performance and predictive tests.

Finally, quality improvement is integrated into each component of the ISD system. Formative evaluation is used to provide in-process feedback about the quality of the training system, including developmental and operational tests. This process takes place during the entire development and design of the training system. Summative evaluation is used to evaluate the completed system functions "on the target audience in an operational environment."[13] This ensures that the training system is meeting the standards determined during the development phase prior to a systemwide rollout. After the training has been fully implemented, operational evaluation continues for the entirety of the training's use. Comprised of internal and external evaluations, the system is continuously monitored to ensure that the system performs effectively throughout its life cycle.

Competency Modeling

The Air Force Handbook 36-2647 provides information material on how and when competency models are planned, built, and executed in support of the USAF.[14] Competency models are used to identify critical knowledge, skills, and abilities that are key to the success of performing a job. This handbook specifies that competencies should be "designed as a way to allow all airmen to build a path towards attaining these behaviors." The competency modeling process is outlined in three phases: development, integration, and implementation.

During the development phase, jobs are targeted, subject-matter experts (SMEs) are identified, and the competency model is developed using qualitative and quantitative methods. These methods include interviewing SMEs, focus groups, and ranking competencies based on the needs of the job and the proficiency needed at entry. The process then moves to the integration phase. This phase involves, as the name suggests, integrating the competency model into a training plan. Learning outcomes for training can be determined using the tasks and behaviors identified through the competency model. The ultimate goal is for the competency model to be integrated into a career field education and training plan (CFETP), which is a "comprehensive education and training document that identifies lifecycle education and training requirements, training support resources, and minimum core task

[12] U.S. Air Force, 2002, p. 10.

[13] U.S. Air Force, 2002, p. 265.

[14] U.S. Air Force, Competency Modeling, Air Force Handbook 36-2647, February 8, 2022.

requirements." Finally, the implementation phase involves running the competency-based training through the CFETP.

This training system will "allow mapping of Developmental Special Experiences (DSE) and courses in MyVector which can provide a catalog of learning opportunities that have been cross-linked with foundational and occupational competencies" provided by the competency model. This can aid efforts to shore up gaps within the training program to ensure a proper linkage with the competencies. Competency maps can be made to track individual progress. Strengths and weaknesses can be determined and then used as individualized training tools to help guide future development.

Airman Comprehensive Assessment Addendum

This form is "designed to guide raters and facilitate discussion when providing constructive feedback to their ratees."[15] Individuals are rated on a scale of proficiency in each competency between "developing," "proficient," "highly proficient," and "outstanding." Each competency is defined, as well as what each proficiency level means in that context. It is specified that raters should use their best judgment when making proficiency-level determinations, and the whole person concept should be applied "relative to the ratee's specific rank, Air Force Specialty Code (AFSC), and assigned duties." For example, "executing the mission" is a major performance area for airman leadership. Three competencies fall under this performance area: job proficiency, initiative, and adaptability.

Initiative is defined as "assesses and takes independent or directed action to complete a task or mission that influences the mission or organization." Under this competency, the "developing" proficiency level is defined as "takes action to provide value to an assigned task or mission." "Proficient" is defined as "takes appropriate action with little or no prompting that provides value to an assigned task or mission." "Highly proficient" is defined as "implements solutions without prompting; encourages others to improve an assigned task and/or mission." "Outstanding" is defined as "anticipates and appropriately takes on complex/critical problems without prompting; inspires others to make improvements in an assigned task or mission." Each competency and proficiency level is defined in a similar manner, increasing in complexity, responsibility, and criticality.

AFSPECWAR Efforts to Develop Nontechnical Attributes

Highlighting a Few AFSPECWAR Practices

As part of our discussions with operational psychologists and subsequent document review, we learned that the Command Control (CCT) Apprentice course has developed a structured approach to providing feedback and resources for developing AFSPECWAR attributes. Although other courses may have implemented some level of feedback, the documentation from the CCT Apprentice course was the most comprehensive and deliberate. We briefly describe these efforts and the instructor qualification course for Special Warfare in the following sections.

---

[15] U.S. Air Force, Airman Comprehensive Assessment Addendum, Air Force Form 724-A.

Feedback Efforts

Although not yet standardized across SWTW, there are example efforts that demonstrate a commitment to providing feedback to trainees about their performance on each of the eight attributes. For example, as a critical part of feedback within the training pipeline, the CCT Apprentice course uses standardized peer assessments to support developmental goals. The peer assessment portion is important because those working together consistently can more accurately evaluate the attributes that others on the team possess. Individuals rate their peers based on the attributes on a five-point Likert-type scale and provide comments as feedback. Efforts have been taken to teach airmen how to give valuable, constructive feedback. One way this is done is through instruction on common cognitive biases that can negatively affect the accuracy of peer assessments. Operational psychologists also teach students about "actionable feedback" that distinguishes between judgments and behaviors. Feedback should be directed toward behavior that is verifiable to outside observers. Making judgments about a person's performance does not provide the peer with action points that can be addressed. Additionally, individuals are instructed to begin each sentence with either "Start," "Continue," or "Stop" so that peers can provide each other with actionable feedback.

Students receive reports based on the peer reviews. These reports consist of average scores for each attribute, supplemented with a spider web graph that compares the self-assessment with the peer assessments. Although some bias is expected in self-ratings, large discrepancies between the self-assessment and peer assessment can indicate a lack of self-awareness among ratees. In addition to ratings, individuals receive a summary of comments made by their peers aligned to each attribute. These comments provide valuable insight into understanding the ratings and offer steps for improvement.

Emotional Intelligence Reports

Students in the CCT Apprentice course also complete self-assessments of emotional intelligence (EI). Using a model of emotional intelligence, individuals gauge themselves on 15 specific EI skills in five different categories: (1) self-perception, (2) self-expression, (3) interpersonal, (4) decisionmaking, and (5) stress management. This assessment is meant to draw connections between emotional and social function and academic, professional, and personal performance.

Each category is a composite of three specific EI skills. Individuals receive scores on each of these skills, a composite score for each category, and a total EI score. Included in the report are explanations for what their score means for each EI skill and strategies one can put into action to increase these skills. A report on the emotional expression skill can give suggestions. For example:

- "The next time you work on a group project, pause and take a breath when the other person speaks. Use this moment to absorb what he/she just said and put aside your inner dialogue."
- "Acknowledge what he/she just said (i.e., address the point or reassure him/her that you will address it later)."
- "Tailor your expression to what you just heard so that your message is well-received."

Developmental Modules and Reading Lists

To support the structured feedback targeting each attribute, some AFSPECWAR courses encourage individuals to reflect more broadly on the scores and feedback they have received. These developmental modules consist of prompts that ask a variety of questions involving participants' thoughts on the feedback they have received and how their strengths and weaknesses could affect them down the line. Particularly emphasized in these modules is thinking about the future, what they do and do not want to be in the future, and what they can do in the present to become what they want to be in the future. The modules posit scenarios regarding an individual's accomplishments in the future:

> Think about your life in the future. Imagine that everything has gone as well as it possibly could. You have worked hard, accomplished many of your life goals, and have "run" your course well. Think of this as the realization of your highest aims.
>
> Who are you in the future? Please describe in detail the person you want to be in life. What character qualities are most important to you? What do you want others to say about you and your accomplishments? Describe what you want to have learned (soft-skills, hard-skills, knowledge, or abilities). What qualities do you exhibit in your relationships (professional and personal)?

A list of reading materials that correspond with the critical competencies was provided to trainees. The material includes mostly nonfiction, self-help books that inform individuals about how specific competencies can be increased.

AFSPECWAR Instructor Qualification Course

The Special Warfare Instructor Qualification Course (IQC) trains AFSPECWAR operators to train AFSPECWAR students. IQC course design and content is founded on recent academic research, including effective instructional practices. Specifically, the content of the course is based on the context and unique needs of Special Warfare (SW), including how to perform technical tasks under extreme and dangerous conditions. Therefore, a focus of IQC is to learn how to provide training in realistic environments. These environments present hazardous and unpredictable conditions, so instructors must have high levels of situational awareness to make appropriate training and safety decisions. With this context in mind, IQC covers a range of topics, including learning principles, complex training scenarios, practice, exercises, motivation of students in stressful situations, professional conduct, group dynamics, and program improvement. IQC was developed to address limitations with the Basic Instructor Course (BIC), which is used to train instructors for non-AFSPECWAR specialties. This training provides an important foundation for instructors to evaluate and provide feedback to students across the AFSPECWAR attributes.

SWTW Training Document Review

As one input to explore AFSPECWAR efforts to develop attributes, we reviewed official training documentation. Having been provided numerous electronic files related to AFSPECWAR's initial training pipeline, we determined which types of documents would be most valuable, developed a coding system, and systematically coded relevant course materials.[16] The goal of this analysis was to

[16] The AFSPECWAR training community provided RAND with over 500 training course documents. Of all the documents reviewed, RAND relied primarily on course charts, lesson plans, and plans of instruction (POIs) to understand each course's structure and content.

identify how often and where attributes and their facets are currently being developed or assessed within the initial training pipeline. Additionally, we wanted to highlight areas where opportunities for further developing attributes may be present, and some similarities and distinctions between courses that may be informative for pipeline and course managers.[17]

This section highlights select findings from the training documentation review. Specifically, we present how AFSPECWAR attributes are potentially represented across the nine AFSPECWAR initial training courses reviewed: Special Warfare Candidate Course (SWCC), Pararescue (PJ) Apprentice, CCT Apprentice, Special Reconnaissance (SR) Apprentice, TACP Apprentice, Air Traffic Control (ATC), Pre-Dive, Open Circuit Dive, and Closed Circuit Dive. Training documents for each course specify course objectives that are the basis for training events in each course. RAND analysts initially coded each course's objectives/events as to whether each attribute facet was directly or indirectly developed, formally assessed, or presented an opportunity for more attribute focus. Given challenges in differentiating direct and indirect development strictly from the objectives/events described in training documentation, we collapsed both into a single "developed" category.

Table 2.1 provides a definition and example objectives/events aligned with attributes that were coded as "developed."

Table 2.1. Example Training Events Developing AFSPECWAR Attributes

Category Definition	Objective/Event	Attribute	Course
Developed: Facet is directly developed or generally reinforced or indirectly developed during objective/event.	Identify relationship of basic facts and state general principles about non-radar separation between arriving and departing aircraft.	Trainability	ATC
	Perform aircraft marshaling.	Teamwork	CCT
	Safely employ grenades.	Stress Tolerance	CCT
	Identify basic facts about intelligence elements for weather and terrain analysis.	Problem Solving	PJ
	Perform hand-to-hand combatives.	Physical Fitness	PJ
	Perform team leader command and control.	Integrity	CCT
	Given a scenario, as a team, perform overland infiltration and exfiltration operations.	Drive	SR
	Compose digital message formats.	Communication	TACP

We also coded objectives/events that were considered to have the potential for developing AFSPECWAR attributes. Coding rules allowed analysts to make judgment calls regarding the potential for further developing attribute facets during any formal training objective or event. However, longer training events with significant time dedicated to student practical application and exercises were most likely to be coded as having an opportunity to expand on current tasks, introduce more challenging conditions, or add a deliberate period of observation with a focus on attributes. This

[17] All the data and findings presented in this section were developed solely from a review of documented course materials, primarily each respective course's course chart, POIs, and lesson plans. This is intended to complement data collection and perspectives from other methods, including interviews and observations described in Chapter 3 of this report.

approach did not have analysts consider periods between formal training events outlined in course documents (priorities of work between tasks, administrative and transportation time, etc.), where additional attribute development or observation opportunities likely also exist. Table 2.2 provides a definition and example objectives/events aligned with attributes that were coded as an "Opportunity."

Table 2.2. Example Opportunities to Develop AFSPECWAR Attributes

Category Definition	Objective or Event	Attribute	Course
Opportunity: Opportunity for deliberate development or evaluation of facet during objective or event.	No objectives/events identified for additional Trainability opportunities.	Trainability	--
	Perform recovery operations.	Teamwork	PJ
	Construct supplemental and field-expedient antennas.	Stress Tolerance	TACP
	Perform Live Fire Day immediate action drills and Live Fire Night immediate action drills.	Problem Solving	SR
	Perform non-GPS land navigation.	Physical Fitness	CCT
	Given a scenario, perform rural reconnaissance patrols.	Integrity	SR
	Perform combat patrols.	Drive	PJ
	Perform personnel control and search procedures.	Communication	PJ

Limitations

Any content analysis of official documentation should be viewed as a starting point for understanding how AFSPECWAR is developing attributes. Training documentation alone does not provide any indication about the quality of training or how the training events are executed. Documents vary in terms of number of objectives and events and level of detail. The coding process applied was based on individual objectives and events that range from a one-hour lecture to multiday exercises, so the results presented here do not account for the overall duration of attribute development. Furthermore, attribute development can occur within a training course but outside of formal training events. Finally, training events may not always follow documentation due to lags in publishing timelines for officially approved policies. Therefore, the linkages presented should be viewed as preliminary and triangulated with inputs from training instructors and students (see later sections in this chapter). For a detailed description of the process used to review and code training documentation, see Appendix B.

Attribute Representation Across the AFSPECWAR Training Pipeline

This section highlights communitywide results from our analysis regarding the degree of attribute representation, differences between courses, and opportunities to inject additional attribute development. Looking across all nine courses, AFSPECWAR attributes appear to be well represented throughout the initial training pipeline.

The proportion of objectives and events throughout the AFSPECWAR initial training pipeline that currently develop specific attributes is shown in Figure 2.1. As expected, Trainability was particularly well represented and primarily coded as "developed" across all courses examined.

Figure 2.1. Attribute Representation Across Nine AFSPECWAR Training Courses

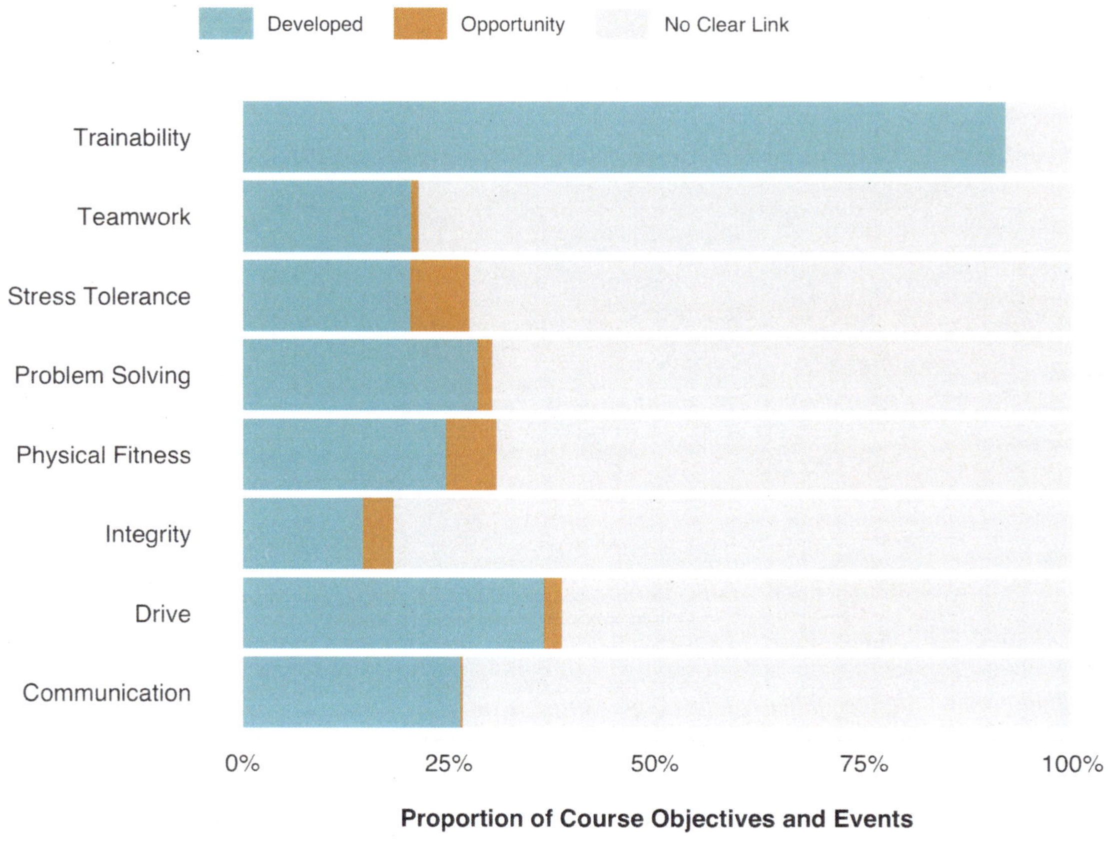

Objectives and events that are administrative in nature were generally the only areas assessed as not having some Trainability-related content.

The Reliability component of Drive was the most frequently coded facet across the nine courses. As a reminder, that facet's description includes "paying attention to details, takes responsibility, and commits to performing high-quality work." Many of the procedural skills taught during AFSPECWAR initial training courses are technical and/or dangerous in nature. As a result, Reliability was often considered embedded or could be developed in numerous training objectives and events.

Problem Solving was also a frequently occurring attribute and one that was well represented across each of its three facets, which is not always the case with other attributes (Communication and Teamwork had specific facets that varied in representation, for example). This demonstrates why facet-level coding was initially helpful to understand the structure of each course. The prevalence of Problem Solving is not surprising, as the objectives and events that comprise each training course originate from individual or small unit tasks—tasks that are often formulated in the context of using some sort of resources to solve problems. Physical Fitness was also found to be common because each course includes physical training along with some physically demanding tasks.

Relatedly, as initial training courses, much of the content is focused on individual performance and knowledge. Thus, while there are tasks that do require collaboration with one or more other students, they appear less frequently than individual tasks, and we see less instances of Teamwork and Communication at this stage in the training pipeline. Generally, most Teamwork-related objectives and events were also coded with at least some Communication facet. However, other tasks related to radio procedures and report generation include elements of Communication but not necessarily Teamwork facets, resulting in slightly more Communication instances than Teamwork.

A Sizable Portion of Each Course Is Structured to Develop Attributes

The relative portion of objectives and events in each course that is structured to develop AFSPECWAR attribute facets is presented in Figure 2.2. Because each course has a different number

Figure 2.2. Opportunities to Develop Attributes Across Initial Training Courses

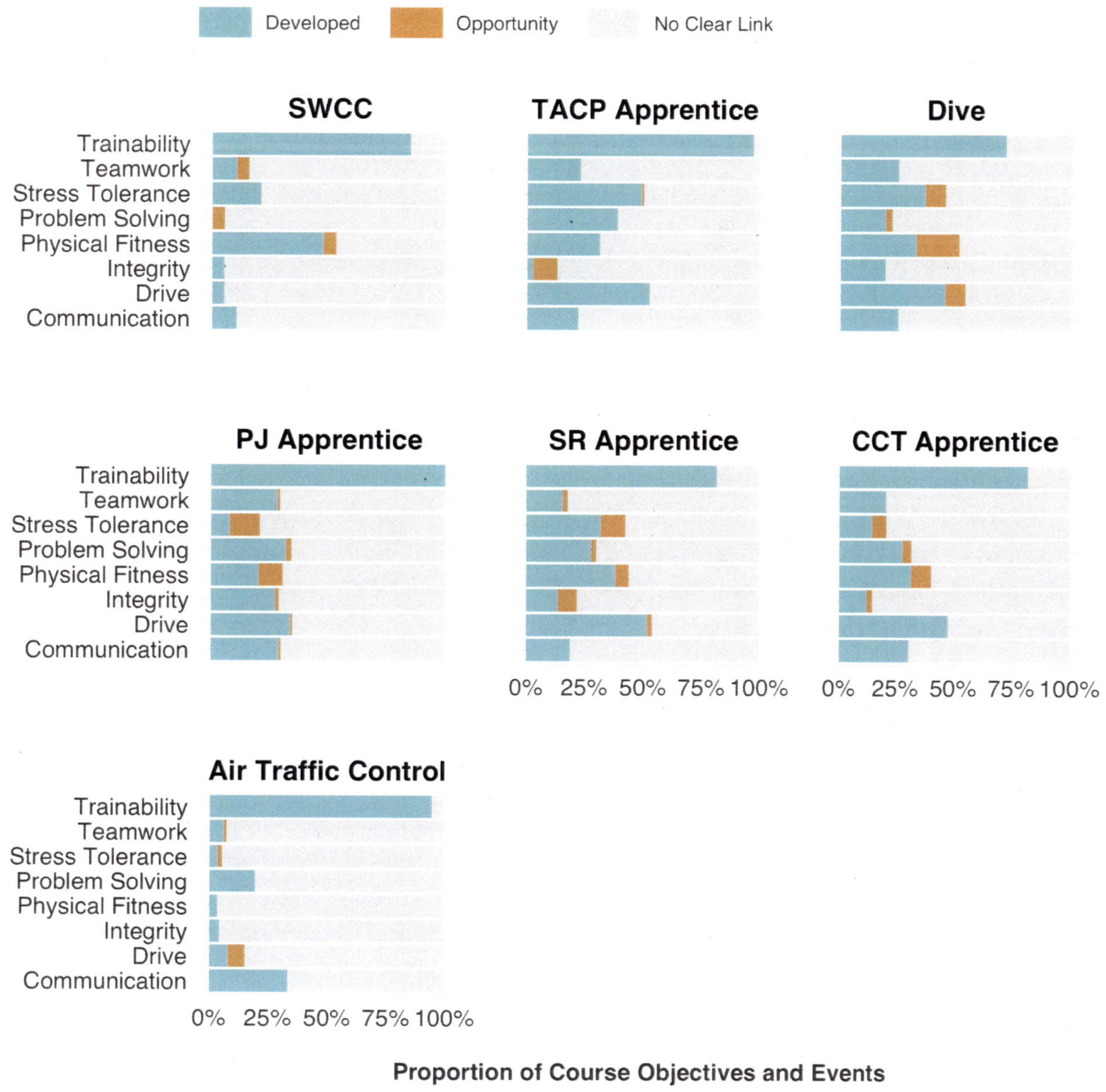

of objectives and events,[18] percentages were calculated by summing the number of objectives and events that were coded as developing any facet. That sum was then divided by the total possible combinations of objectives and events and facets for each course.

Two challenges with interpreting these results at the course level are the wide disparity in how many objectives and events are in each course and the degree to which objectives and events include multiple tasks or performance steps versus shorter, more discrete events. A course with a larger number of objectives and events that are each very discrete and more likely to address fewer attributes at once will have a lower percentage of the course (as measured in terms of the total possible combinations of objectives and events) representing attributes than the converse. For example, the three entry-level dive courses (all combined in Figure 2.2) collectively have 35 objectives and events that were coded. Many obviously involve activities in a pool or open water. Therefore, we observed a relatively large percentage of dive training that develops Physical Fitness, Stress Tolerance, and Drive. Additionally, a single objective/event in Pre-Dive includes 129 hours of various water skills tasks that were tied to different attributes but all coded for the single objective/event. Alternatively, the PJ Apprentice course has over 200 objectives and events with some as discrete as "perform immobilization and splinting procedures." Thus, the overall percentage of objective and event and attribute combinations coded as developing any particular attribute should be expected to be smaller.

Of the 19 objectives and events coded for SWCC, many have the primary goal of developing students' strength and endurance, while other more job-specific tasks are reserved for AFSC-producing courses. Thus, it is not surprising that a much greater proportion of objectives and events during SWCC emphasize Physical Fitness over other attributes. Physical Fitness is still represented fairly frequently as a proportion of course objectives and events in other courses as well but generally more balanced with other attributes.

Another finding from a course-level examination was that ATC had the lowest portion of its course related to AFSPECWAR attributes. While containing over 100 objectives and events, a large majority of training events described in the course documents are conducted as lectures and demonstrations. Technical classroom instruction was found to contain relatively fewer examples of attribute development than student practical application or exercises, which are much more common relative to the overall number of objectives and events in other AFSC-producing courses.

Some Objectives and Events Support Multiple Attributes

Many of the objectives and events assessed were determined to develop multiple attributes at the same time. Figure 2.3 shows how often attributes are developed in isolation or in combination with other attributes. Physical Fitness, for example, has 44 instances where it is the only attribute being developed due to numerous objectives/events dedicated to physical training alone. Many objectives and events across courses also require students to apply resources to developing or implementing solutions to accomplish a short task (Problem Solving) or employ radios and report formats (Communication) that do not address other attributes. The Reliability component of Drive was also commonly coded by itself because many shorter tasks with no obvious ties to other attributes require students to pay close attention to the procedures they are performing. Additionally, the Obligation

[18] For a comparison of course length and the number of objectives/events per course, see Figure B.3 in Appendix B.

Figure 2.3. Number of Objectives and Events Developing Different Combinations of Attributes

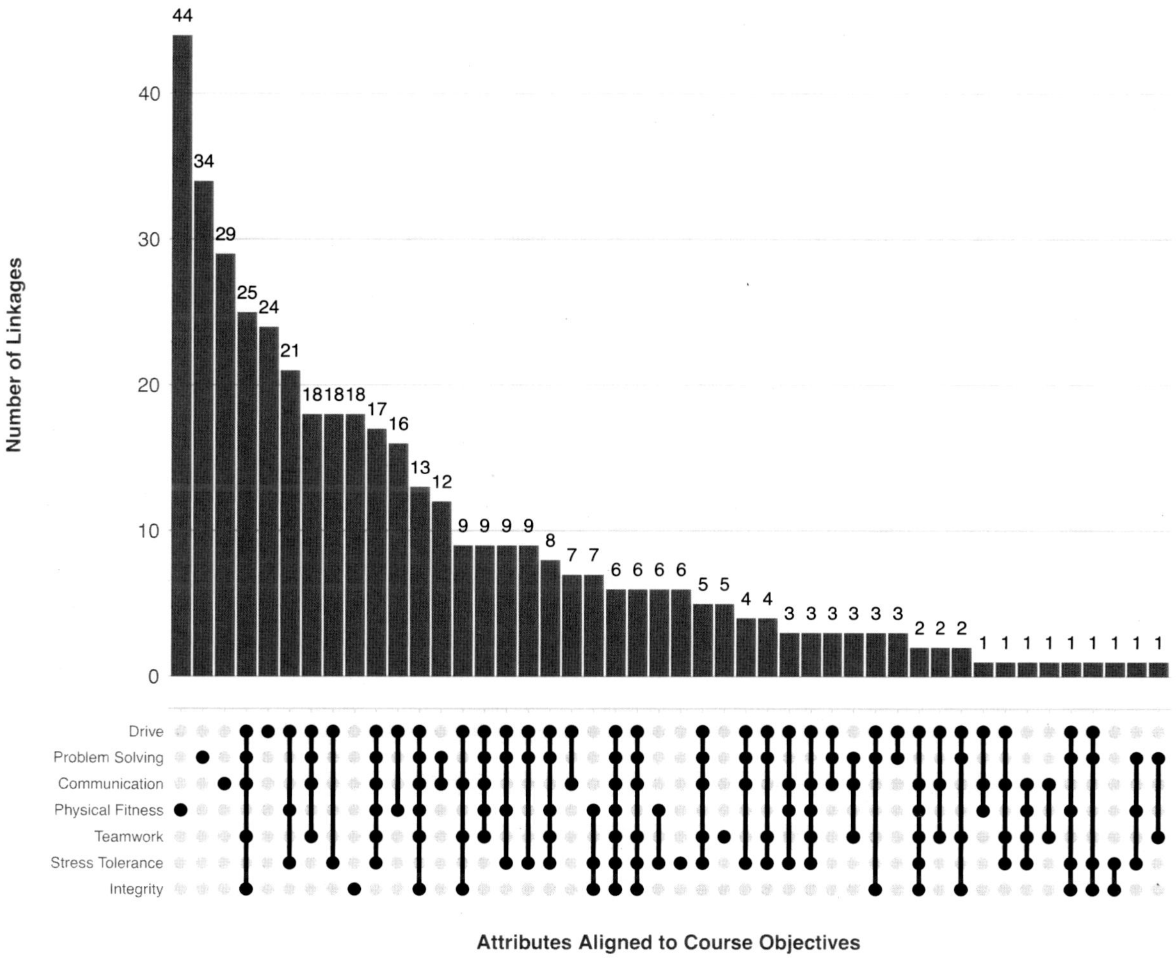

NOTE: Due to the high number of interactions, Trainability is excluded from the figure.

facet of Integrity was coded for each formal examination, given the expectation that students do not cheat on tests.

It was more common, however, that more than one attribute is developed by a single objective or event. In fact, every attribute intersects with every other attribute in at least one objective or event across the courses. One of the most common intersections between attributes was Teamwork, Communication, and Problem Solving. As noted earlier, most objectives and events conducted as a small team include an inherent expectation that students communicate well with their teammates. Many of those tasks also include some element of Problem Solving that requires working together as a team to be successful. For example, numerous patrolling techniques, recovery operations, and methods for determining aircraft control measures all require students to understand environmental conditions, serve in specific roles, and communicate within or outside their assigned team.

Stress Tolerance, Physical Fitness, and Drive were also commonly found to be developed in the same objectives and events. This is largely attributed to longer duration tasks or exercises that require Persistence, which is a facet of Drive. Whether moving long distances across land or in water or other physically demanding tasks such as recovery operations, longer duration tasks were also likely to be coded for developing Physical Fitness. Additionally, the Performance (Under Stress) facet of Stress Tolerance is represented in longer tasks, particularly when environmental conditions are challenging. Thus, we would expect there to be frequent overlap across these attributes.

Opportunities to Further Develop Attributes Across All Courses

We also coded objectives and events that were considered to have the potential for further developing AFSPECWAR attributes. Training events with significant time dedicated to student practical application and exercises were most likely to be coded as having an opportunity to expand on current tasks, introduce more challenging conditions, or add a deliberate period of attribute observation. Although the percentage of overall course objectives and events with additional opportunities is relatively small (Figure 2.2),[19] the raw number of opportunities varies across courses and by attribute primarily due to course length.

The two attributes found to have the most opportunity for further development across all their facets were Physical Fitness and Stress Tolerance, despite already being fairly well developed through a number of training events. These results reflect a training environment that can be easily modified to vary training conditions (e.g., time pressure) and physical demands (e.g., distance to move on foot). It was easier to imagine how those conditions could be made more challenging within existing objectives and events.

The only facet scored as having more opportunities to develop than currently being developed was Courage, which is a facet of Integrity. There are relatively few training events that pose real physical danger to overcome or require students to demonstrate Courage. However, some team tasks could be designed to force students to make difficult decisions related to what is best for the mission and team at the expense of the individual. While those objectives and events were often coded as developing Teamwork, "sacrificing for the greater good" is a component of Courage, and it was not always clear if team tasks presented this dynamic; thus, they were coded as potential opportunities.

Given the nature of initial training and its emphasis on individual tasks, there are fewer opportunities to develop Communication and Teamwork. Additionally, when objectives and tasks require interaction among students, Communication and the Roles facet of Teamwork were coded as being developed, leaving few existing objectives and events with additional team-oriented opportunities.

Formal Evaluations Are Mostly Linked to Trainability and Physical Fitness Attributes

We also attempted to code training documents for attributes represented in Progress Evaluations/ Progress Checks. Except for SWCC, every course includes at least one objective or event that could be considered a formal evaluation, labeled differently across courses as measurements, tests, or evaluations. These are generally represented by three types of tests: (1) written, (2) physical fitness,

---

[19] The coding rules allowed analysts to make judgment calls about where additional opportunities for attribute development existed based on their understanding of how each objective or event is executed, as described in the training documents. However, each objective or event was only assigned one code for each facet, so if an objective or event was determined to already develop a facet, it could not also be coded as an opportunity. Additionally, opportunities were only coded for within existing objectives/ events because we did not consider the addition of new objectives/events dedicated to any given attribute.

and (3) operational skills. However, most objective or event titles that included terms such as measurement, test, or evaluation and those with descriptions that include instructions for conducting Progress Evaluations/Progress Checks often were not labeled in the training documentation as Progress Evaluations/Progress Checks as a training method. Many were instead labeled with a Performance practice training method or no training method at all. This greatly complicated attempts at standardizing coding rules for formal evaluations.

Additionally, there are very few formal evaluations relative to the total number of objectives and events constituting these courses, even with the most liberal interpretation of objectives and events not labeled as formal evaluations. Broadly speaking, the AFSC-producing courses have a written exam at the conclusion of training blocks (although some have two, and many have none listed). Some have multiple physical fitness tests throughout the course; others do not. And other than two dive courses, the SR and CCT Apprentice courses are the only ones with formally defined operational skills tests.

As a result, there are relatively few examples of direct attribute evaluations. That said, tests measuring physical fitness have clear linkages to Physical Fitness facets, and written exams were all considered to evaluate the Learning Content facet of Trainability. Written exams were also considered as developing the Obligation facet of Integrity, given the expectation that students do not cheat on tests. Trainability's Learning Procedures facet was more likely found in the few operational skills tests. Those tests also pose the greatest opportunities to evaluate attributes other than Physical Fitness and Trainability, but again, they are not included in the training documentation for most courses.

Document Review Conclusions

As noted previously, the training document content analysis is a first step in identifying how courses are structured to develop AFSPECWAR attributes. Overall, this analysis suggests the following preliminary conclusions:

- DAF has several related ongoing efforts to develop nontechnical attributes (i.e., foundational competencies).
- AFSPECWAR attributes are quite common, albeit unevenly developed, across initial skills training courses.
  - Trainability is widely represented since the purpose of these entry-level courses is to impart knowledge and skills, the definitions of Learning Content and Learning Procedures facets. Objectives and events that are administrative in nature were generally the only areas assessed as not having some Trainability-related content.
  - Drive is the most common attribute developed other than Trainability. Although formal entry-level training does not present many opportunities for students to display Initiative (within the defined objectives and events), opportunities may exist outside the classroom and prior to or following events. However, many tasks require paying attention to detail and taking responsibility for outcomes (Reliability), and longer tasks and exercises can develop or display students' Persistence.
  - Problem Solving is linked to many objectives and events across the courses. Most skill sets developed during initial training require that students understand the situation and

objective, develop or recall an appropriate course of action, and implement a solution. Thus, all three Problem Solving facets are fairly well represented across courses.

- Teamwork presents fewer opportunities for development in initial skills training because of the emphasis on individual proficiency. However, AFSPECWAR tasks that are performed in small teams develop individual roles and responsibilities for operating and coordinating with others to complete an objective. Thus, numerous objectives and events are conducted "as a team," and Teamwork is represented, primarily through the Roles facet.
- Communication, in some form, is almost always required to complete tasks where Teamwork is involved. Thus, we find these two attributes closely connected throughout training. Communication is slightly more common in some courses given dedicated training on radio procedures and standardized reporting formats.
- Integrity was not linked to many objectives or events, and it would be difficult to structure events to elicit Integrity without creating ethical and safety issues for students. That said, the Obligation facet was considered important during written exams (not cheating) and in a few specific cases where the Roles facet of Trainability and the Reliability facet of Drive intersected, especially in situations where avoiding misconduct when fulfilling personal duties is particularly acute (e.g., medical interventions).
- Stress Tolerance is developed when the nature of the task and conditions are particularly demanding, as seen through combatives and nighttime live fire tasks, for example. Other tasks, however, are taught and practiced in ways that deliberately do not introduce additional stress. This is because adding stressful conditions would likely detract from initial skills development, particularly when tasks are novel, technical, or complex.
- Physical Fitness is covered through many physically demanding events, training, and official fitness tests reflecting the physically demanding nature of these career fields.

• Opportunities to develop attributes were mostly associated with Stress Tolerance and Physical Fitness because the environmental conditions under which tasks are performed are relatively simple to adjust in those domains, if desired. Even simple radio procedures and marksmanship can be designed to incorporate these two attributes if the tasks are conducted with time limitations, in the dark, around loud noises, and following a run and grappling session.

Chapter 3

Training Needs Analysis for SWTW Nontechnical Attributes

As demonstrated through our document review, students are exposed to nontechnical AFSPECWAR attributes throughout the training pipeline. In this chapter, we review two streams of information to form the basis of our needs analysis for SWTW. First, we review perspectives from students, leadership, and instructors on how much the current training pipeline emphasizes attribute development. Next, we consider training performance evaluation data that we collected as part of a related project for AFSPECWAR.[20] Specifically, we coded which, if any, of the AFSECWAR attributes were involved in rater comments about poor-performing students.

Student Perspectives on Attribute Development

Between November 2021 and March 2022, 32 focus group sessions were conducted with AFSPECWAR students. Participants were comprised of both enlisted airmen and officers who had already completed Basic Military Training (BMT) and were attending one of the following courses:

- ATC
- CCT/STO Apprentice
- PJ/CRO (Combat Rescue Officer) Apprentice
- SR Apprentice
- Special Tactics Training Squadron (STTS)
- TACP Apprentice
- TACP Initial Certification Course (ICC) Phase I
- TACP ICC Phase II

During the focus groups, students ($N = 228$) completed a brief survey asking them to rate how much training or development on each AFSPECWAR attribute is provided in their current training course. Ratings were provided on a five-point scale:

1 = Not at all
2 = Very little
3 = Some
4 = A fair amount
5 = A great extent.

[20] Krueger et al., 2024.

Student Ratings

Except for Teamwork in the TACP ICC Phase II, students rated all attributes as being emphasized to "Some" degree. The apprentice courses that were consistently rated a 4 or higher (i.e., emphasized "A fair amount" or more) were Trainability, Teamwork, Stress Tolerance, and Drive. The least emphasis across courses was Communication, with an overall average of 3.87 (Table 3.1). The distribution of student ratings across courses and attributes is illustrated in Figure 3.1.

Table 3.1. Average Student Emphasis Ratings by Course

Attribute	ATC (*N* = 11)	CCT/ STO Appr. (*N* = 25)	PJ/ CRO Appr. (*N* = 45)	SR Appr. (*N* = 9)	STTS (*N* = 21)	TACP Appr. (*N* = 68)	TACP ICC Phase I (*N* = 33)	TACP ICC Phase II (*N* = 16)
Trainability	3.82	4.48	4.34	4.67	4.32	4.40	4.19	4.33
Teamwork	3.30	4.56	4.07	4.67	4.76	4.39	3.88	2.67
Stress Tolerance	3.50	4.08	4.18	4.44	4.00	4.56	3.64	3.93
Problem Solving	3.50	4.28	4.16	3.89	4.38	3.90	4.39	4.44
Physical Fitness	4.00	4.24	3.27	4.33	4.48	4.50	3.41	3.67
Integrity	4.00	4.04	3.92	4.33	4.38	4.33	3.78	3.80
Drive	3.30	4.36	4.07	4.78	4.52	4.27	3.75	4.13
Communication	3.09	4.12	3.73	3.89	3.86	3.98	3.91	3.80

NOTE: *N* represents the number of students providing at least one attribute emphasis rating.

Figure 3.1. Distribution of Student Ratings of Attribute Emphasis by Course

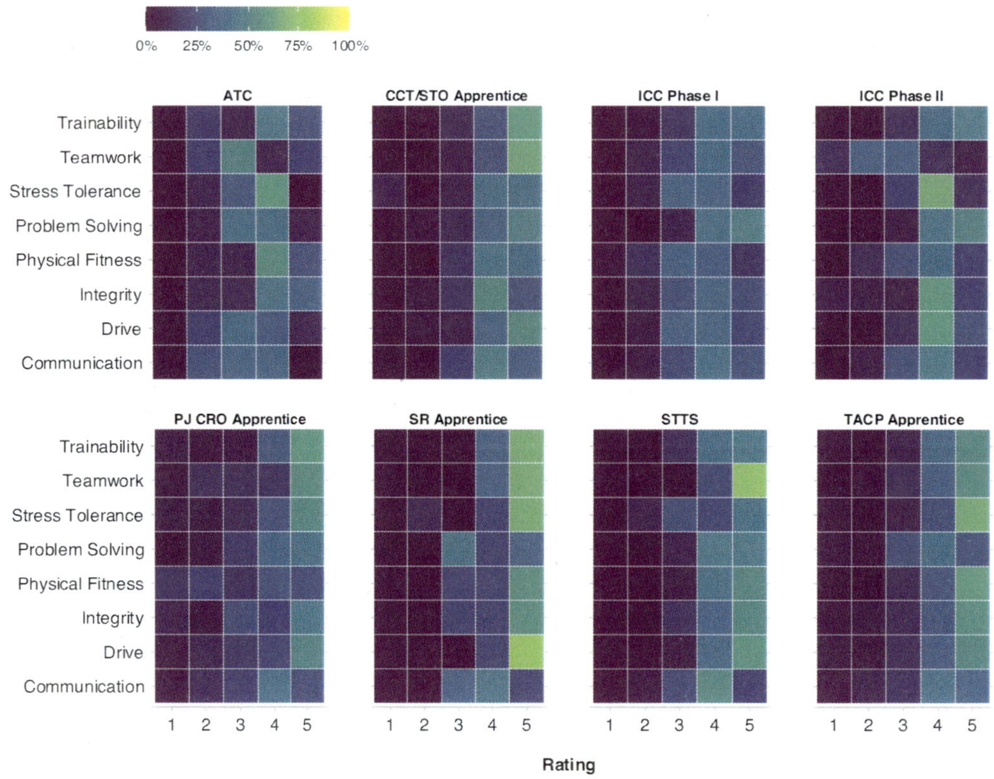

Student Focus Group Discussions

A consensus discussed by most of the student participants was that most were familiarized with the attributes during the Special Warfare Preparatory Course (also referred to as Prep)[21] and assessment and selection (A&S). Beyond Prep, and during their follow-on training courses, students indicated that attributes were seldom discussed explicitly.

> It just was mainly emphasized in Selection. (Trainees CCT STO SR STTS)
>
> After A&S, I would say it was hit on minimally. (Trainees CCT STO SR STTS)
>
> Yeah, that was in Prep. I remember they talked about. (Trainees TACP Apprentice)
>
> I've been exposed to—I went through MARSOC [Marine Forces Special Operations Command] Assessment and Selection, and this is the only branch that does Assessment and Selection on attributes versus skills. And this is something that I talked [about] with some of the staff as well, that BMT doesn't set up guys for success for this type of selection. Prep—essentially, one of the staff members was saying—Prep is BMT 2.0. (Trainees CCT STO ATC)

When attributes were discussed in courses following Prep and A&S, comments indicate that feedback about them was vague or nonexistent.

> So, [the psychologist will] give you a printout and they'll overlay basically how you evaluate yourself versus how everybody else is perceiving you. And then he'll put all the comments . . . about you in that printout based on the different sections of integrity or communication or whatever the set is. He doesn't go over it with [you] individually. (CCT STO SR Apprentice)
>
> A lot of the things that they're looking for at selection weren't really taught before selection. (CCT STO SR Apprentice)

There were some instances when attributes were discussed specifically as having been directly or indirectly developed in the pipeline. Some noted that the training courses prepare individuals overall across several of the attributes. Others discussed specific attributes and when they were trained.

> Yeah. I think all of these—all of these—you develop all of these throughout the pipeline. You're working on yourself continuously, day in, day out, so by doing that, all of these [attributes] should fall into place. (CCT STO SR STTS)
>
> I [rated] the bigger core attributes, 5s, and Problem Solving because if you look back to our sims and our mission building for all of our fields and whatnot. It's all about, "Here are some tactical problems that you need to solve," and you work as a team or you work individually, but usually as a team in . . . like fields. How you're going to solve the problem, how you're going to execute this mission, and then for the sims, it's . . . individual-based because each sim is individual, so how are you going to solve the tactical problem of that sim?. . . Trainability—because they give you this course [that] is four months of a firehose of information. They're teaching you—we

---

[21] This course has been substantially revised and is now named the Special Warfare Candidate Course (SWCC) (implemented in spring 2022). As of June 2022, officer candidates do not attend SWCC.

> probably have, like, 60 or 70 different PowerPoints that you had to know and gain information from. So learning the content . . . you have to acquire that knowledge, and then you have to therefore apply that knowledge in those tactical problems, in those sims, in those missions. Sometimes you learned it the day before, or the day of, and then you had to execute that the next day or that day. (TACP FTU Phase I)

Leadership and Instructor Perspectives on Attribute Development

Between December 2021 and March 2022, ten discussion sessions were conducted with AFSPECWAR training course leadership and instructors. Across these sessions, participants discussed their efforts on training nontechnical attributes, the effectiveness of those efforts, and potential areas for improvement. Instructors and course leadership represented the following courses:

- CCT/STO/SR Apprentice
- CCT/STO/SR STTS
- PJ/CRO Apprentice
- TACP Apprentice
- TACP ICC Phase I
- TACP ICC Phase II

Instructors and course leadership (N = 37) also completed the same survey asking them to rate how much training or development on each AFSPECWAR attribute is provided in their current training course. We first present findings from the survey ratings, followed by a summary of discussion topics.

Leadership and Instructor Ratings

Although there were fewer instructors across courses, most indicated that the attributes were emphasized during their course. Averaging across courses, the least emphasized attributes were Communication (3.72), Teamwork (3.81), and Integrity (3.87). Analyses by training course suggested slightly different patterns depending on the attribute and course (Table 3.2 and Figure 3.2). For example, TACP FTU Phase I emphasizes Communication. Another notable difference is that Problem Solving was emphasized less by the TACP Apprentice course compared with other AFSC apprentice courses.

Table 3.2. Average Instructor Emphasis Ratings by Course

Attribute	CCT/STO/SR Apprentice (*N* = 5)	PJ/CRO Apprentice (*N* = 5)	TACP Apprentice (*N* = 8)	TACP ICC Phase I (*N* = 8)	TACP ICC Phase II (*N* = 11)
Trainability	4.80	3.25	3.50	4.50	4.20
Teamwork	4.40	4.00	3.62	4.12	3.40
Stress Tolerance	4.00	4.20	3.86	3.38	4.18
Problem Solving	4.20	4.50	2.86	4.25	4.40
Physical Fitness	4.60	3.33	4.00	4.75	4.60
Integrity	4.25	3.75	4.29	3.94	3.30
Drive	4.00	4.25	3.57	4.31	4.00
Communication	4.00	3.12	2.88	4.75	3.64

NOTE: STTS and ATC are not shown due to small sample sizes (i.e., fewer than five instructor respondents). *N* represents the number of instructors providing at least one attribute emphasis rating.

Figure 3.2. Distribution of Instructor Ratings of Attribute Emphasis by Course

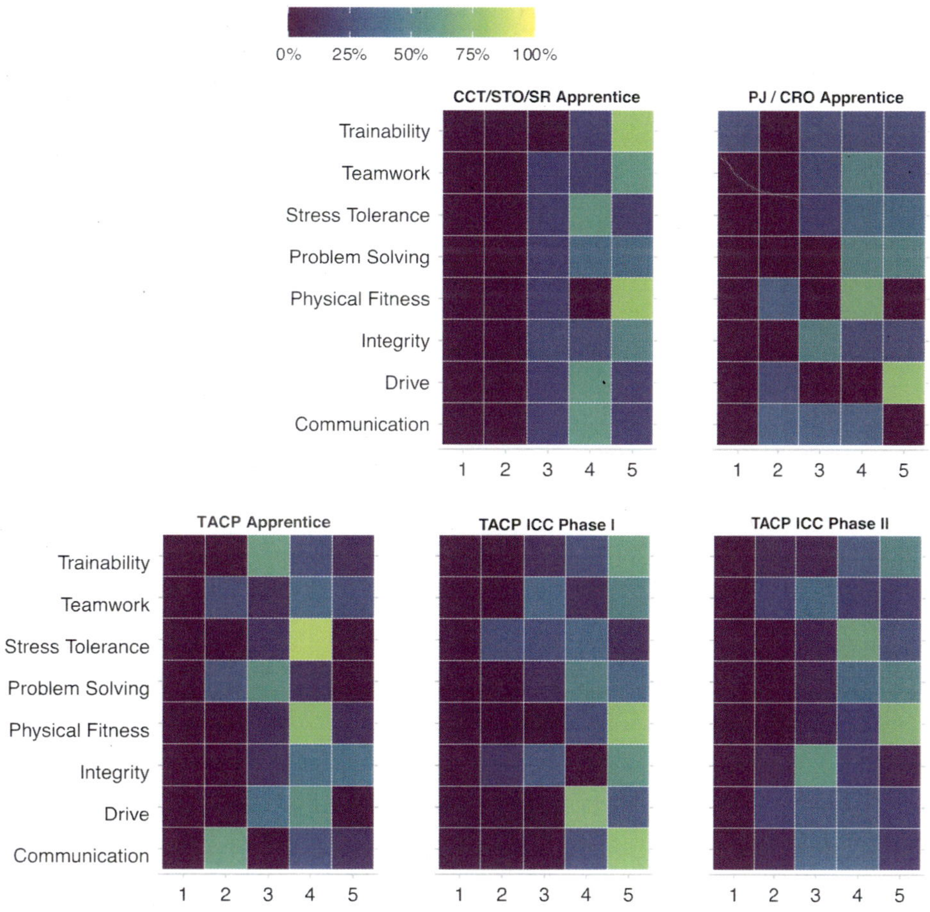

NOTE: STTS and ATC are not shown due to small sample sizes (i.e., fewer than five instructor respondents). Brighter colors in the heatmap indicate a higher proportion of respondents indicating that level of emphasis.

In addition to these quantitative ratings, students and instructors were provided with an opportunity to add comments for each attribute rating. Although most respondents did not provide comments, a few instructor comments suggested that the emphasis on training for certain attributes such as Problem Solving should be higher than it currently is. However, the length of the course and availability of training resources (i.e., instructors) are important considerations in how well these attributes can be developed. For some attributes (e.g., Physical Fitness), content experts and trained professionals are available to provided dedicated support. Finally, a few comments noted the relevance of full mission profiles (FMPs) to developing multiple attributes including Teamwork, Communication, and Problem Solving.

Leadership and Instructor Group Discussions

Participants discussed the ways that attributes were trained, as well as perceived opportunities for attribute development. The eight attributes are listed in Table 3.3 with a count of the number of excerpts in which they occurred across discussions.

Table 3.3. Attributes Discussed in Excerpts from Leadership and Instructor Discussions

Attribute	Excerpts
Drive	57
Teamwork	56
Stress Tolerance	51
Problem Solving	47
Integrity	31
Physical Training	29
Communication	26
Trainability	22

In discussing the attributes, participants widely corroborated their relevance to the development of AFSPECWAR operators. While these attributes are desired, targeted training focused on developing them exists for some but not others during training courses. Regardless of whether they are developed, comments from the interviews indicated that all eight of the attributes are observable in the behaviors of students throughout the courses.

Some Attributes Are Expected but Not Deliberately Developed

With certain attributes, trainees may be expected to exhibit these attributes, at least to some degree, when they enter the training pipeline. Students who do not display these attributes are viewed as not performing at the expected level. Comments throughout the discussions indicated that while many activities provide opportunities to demonstrate these attributes, focus on their development is limited. Some attributes are not developed because there is a perception that they should already be present in students who have been prescreened and/or selected. Others are seen as more stable attributes that are difficult or unlikely to be effectively trained. Attributes in this category include Integrity, Drive, Stress Tolerance, Teamwork, and Trainability.

Integrity

Across the interviews, Integrity was discussed as an important characteristic, with students having opportunities to demonstrate it. However, there were relatively few examples of students receiving deliberate training to develop this attribute. In general, leadership expects students to have it, and there is little tolerance for a perceived lack of it.

> We can teach somebody how to do pull-ups. We can't teach somebody accountability. (Instructor CCT STO SR Apprentice)

> Behaves in an honest, fair, and ethical manner, fulfilling duties, following rules, avoiding misconduct and abuse. We don't teach that. We expect it. (Leadership PJ/CRO Apprentice)

> I don't know if we formally develop and train our trainees for integrity. In my mind, character and integrity is something you have or you don't. (Leadership TACP ICC Phase I)

When students exhibit a lack of Integrity, they may receive both feedback on their behavior and corrective action, along with facing the natural consequences of their actions.

> And I think 360s get back at this again, because the candidates have more interactions with each other, and they can probably tell you. (Leadership PJ/CRO Apprentice)

> I think the course naturally [identifies concerns]—like UCMJ [Uniform Code of Military Justice] and other stuff. . . . So, if a guy goes out on the weekend and gets like a DUI, that will come with certain things, so that will highlight Integrity. (Leadership TACP Apprentice)

> The only development we would do would be through corrective action, to be honest, where they learn from that mistake and don't do it again. (Leadership TACP ICC Phase I)

Some deliberate development occurs in a limited way during ethics-based discussions and scenarios geared toward officers.

> The only time I emphasize it is actually in my initial expectations administratively when I get CROs here. . . . So, we cover that pre-course. (Leadership PJ/CRO Apprentice)

> Specifically, with our STO and STO candidates, . . . every couple months, we have tried to run an ethical leadership type, vignette-based scenario where . . . we're faced with two bad choices. Let's talk through the calculus about why this decision was made over this, and then some of the constraints and restraints going into that decision. Is there a specific class on it? No. (Leadership CCT STO SR STTS)

Participants suggested some things that could be incorporated into courses to promote Integrity among students. Integrity-focused mentorship that is consistent could be standardized and promoted within the context of AFSC and in the personal lives of students outside work.

> And in order to better develop integrity, I think mentorship, just incorporating mentorship wherever, can be useful. (Leadership PJ/CRO Apprentice)

> We have these little quirks with the individual instructors like, oh, you don't have to do that. Don't worry about that. Or you're doing this. And it may not be big things, but it eventually leads to big things after they become . . . involved in that kind of thinking . . . the instructors with the students. (Leadership TACP FTU Phase I)

How Students Are Doing on Integrity

Integrity violations do sometimes occur. When encountered, such behaviors are clearly noted by cadre and leadership.

> Recently there was a student who lost their weapon out in the field, didn't report it up. They got to a point later on in the day and the student tried to say that the instructors took it from them. The instructors went back and followed where they went on their GPS track and found it. They had left it somewhere. So that was obviously highlighted, so there are issues with integrity. (Leadership TACP Apprentice)

> I gave a specific brief, a brief that you guys had, same thing every student has had for years about, okay, you can get your boots now. You've reached that level. Here's how you prepare your boots. And half the class disregarded that and went downtown and paid somebody to do that for them. And I was livid because that's an integrity violation, failure to follow, and they're just looking for the easy way out. (Instructor CCT STO SR Apprentice)

Drive

Throughout the discussions, Drive was the attribute mentioned most frequently across the coded excerpts. Participants regarded Drive as indicative of trainee potential and commented on regularly observing Drive, and in some cases a lack of Drive, among trainees. This attribute is generally viewed as necessary for success, and there is a common perception that this attribute will be present in high-performing students and is crucial for success. Opportunities for trainees to demonstrate Drive are commonplace across the courses, but specific experiences to develop Drive are not.

> The rucking, everyone looks at rucking as a physical activity. And, sure, it's physically challenging, but it's really a mental thing. It's just kind of left foot, right foot, and just get from start point to finish point, and just kind of push through it. It's not going to be fun for anyone. It's going to be probably a little painful. (Leadership TACP Apprentice)

> We see it in a couple different aspects. Initiative, like I said, we don't have enough time to teach these guys. So, you end up seeing who studied enough, who practiced on their own, because when they don't, they fail. (Leadership PJ/CRO Apprentice)

> Yeah, if you don't maintain a high level of Drive, you're not going to be successful, so that's important. But I don't know if we are training that appropriately. (Leadership TACP FTU Phase I)

Although not deliberately developed throughout the courses, leadership pointed out that Drive can potentially be encouraged through feedback and mentorship from peers and cadre.

> Like an individual may just be skirting by, but then, when it's being observed or noticed by an instructor, . . . then we get involved or we basically pull them aside and

say, hey, we're seeing this. This is what your teammates are saying. Most of the time, these guys will change their course or path, or they'll improve. We hope they improve. So that would be, I guess, one way to help develop that Drive is [to say] hey, we're identifying this, and then hopefully it makes them click. Alright, I'm not performing well, or my teammates are saying this about me. (Leadership TACP ICC Phase I)

It was also noted that there is potentially an opportunity for some realignment in the way coaches[22] support students and approach the development of Drive during the courses. The comments below indicate a perception that the current approach may diminish the need for students to be adequately driven to succeed in the course.

> It's no longer on the student to make it through, and that's because HP [human performance] thinks it's their role to get them through. And none of the HP staff have ever done this. They purposely took HP staff—as far as the strength coaches, at least—they purposely took strength coaches with a college training background for the most part. And that's so much different than what we're training for. (Instructor CCT STO SR Apprentice)

> The HP staff will just follow them around with the truck full of niceties. Like, they have ice baths and cooling towels and DripDrop, and all this stuff. And . . . they're like, "Oh, he looks like he's getting a little dehydrated. Here's some stuff," or "He looks like he's getting a little warm. Here's a cooling towel." You're not developing any sort of toughness in these guys doing that. And basically, since I've been here, I've seen nothing but that from the HP side, where they just coddle the students as soon as things get tough. (Instructor CCT STO SR Apprentice)

How Students Are Doing on Drive

Comments suggest that Drive is commonly observed in students and perceived to be directly related to their success in the courses.

> They were here years and years ago talking about what makes a PJ? What do you guys have in common? They were trying to figure out our common denominator. And one of the guys literally was like, "You want to know what we have in common?" And they're like, "What is it?" And he's like, "We didn't fucking quit. That's what we have in common." And that's literally it. (Leadership PJ/CRO Apprentice)

> Again, from our perspective, all the students want to be here. They show an incredible Drive. (Leadership PJ/CRO Apprentice)

> And we see it in the gym. First thing in the morning when guys go to the gym, you can see who's making the effort to improve themselves in the gym. There will be days in training where the schedule doesn't necessarily dictate or allow them to work out in the morning, but you do see a core group of individuals that still come in and make the effort to work out after they train that day. (Leadership TACP ICC Phase I)

---

[22] Coaches are not instructors and are not typically AFSPECWAR operators. Instead, they are content experts in their respective areas of expertise.

> No, overall, everybody is extremely excited to be out of the pipeline and to start focusing on what they're going to do at their line unit. Like that's the big thing for them. The perception is I need to be really good when I get here. (Leadership CCT STO SR STTS)

Some comments about trainees not having a consistent level of Drive throughout a course suggest that students are figuring out at what point during the course they can get away with decreased effort. While this behavior may diminish cadre and leadership perceptions of a particular student's performance, students can decrease effort during certain activities with no negative consequences.

> Guys just don't want to bear down and just get it done with. They're trying to find an easy way to get everything accomplished and take shortcuts on just about everything now. And that is a theme that's been going on for probably about—since I've been here. (Instructor CCT STO SR Apprentice)

> So, are they giving it their all for everything they're doing? Absolutely not. Like 100 percent they are not trying in this course. And they have figured out when and where they can let go. I don't know how they figured it out, but they did. And we are unable to adjust this course to make sure that we hold them accountable to it. (Leadership TACP Apprentice)

Stress Tolerance

All courses in the AFSPECWAR training pipeline have stressful conditions at various points. However, Stress Tolerance is generally not deliberately developed by each course, and to some degree, it is an attribute that students are expected to already possess.

> I don't think at any point, though, we're making a conscious effort to develop stress tolerance. I mean, it's a side effect . . . of what we do here. (Leadership TACP FTU Phase I)

> In coms [communications] block of instruction, okay, we're going to analyze your ability to tolerate stress. That's not it. It's are you able to plug it on the radio? Here's the time block. (Leadership CCT STO SR STTS)

> MODERATOR: Do you expect them to have the stress tolerance that's needed and that you no longer need to develop that?
> RESPONDENT: Absolutely. . . . So that's what A&S is supposed to provide. (Leadership CCT STO SR STTS)

Common conditions that present opportunities for students to demonstrate Stress Tolerance include regular evaluations and progress checks, high physical demands, high workload, and the natural consequences of failure.

> There are a lot of progress checks in this course, a ton of them. And the majority are academic, but that is stressful. (Leadership TACP Apprentice)

> Their task mission is probably stressful because they want to perform. And our instructors are going to apply an appropriate amount of stress, but it's not like at a "3" level. We're not yelling at them; we're not—inducing stress daily. (Leadership TACP ICC Phase I)

> And on top of everything, there's the fear of reclassification. So, in the event that they don't pass, meet the mark for everything, then they get rolled into a different career field. So that's always in the back of their mind as well. (Leadership TACP ICC Phase I)

Comments indicate that incorporating an appropriate level of stress is important to make sure that students can succeed while still experiencing appropriate stress levels. There is an understanding that courses should not be so stressful that learning is stifled; however, students should be able to apply skills learned under stressful conditions.

> I'll tell you what we do now—and this probably applies both for Problem Solving and for Stress Tolerance—[we are] trying to strike the balance between how much do you firehose them with challenges and information versus how easy do you go. And the side that we default to is the firehose, because generally speaking, the harder you push them, the better they do. (Leadership PJ/CRO Apprentice)

> But performance under stress, I gave it a [rating of] 3 because we do have the ability to help them along during the hardest parts of the task in most cases. We are able to give them a helping hand, give them a reminder, so we don't let them just flounder. (Leadership PJ/CRO Apprentice)

Participants described some opportunities for developing Stress Tolerance by applying incremental physical and mental stressors during training and providing feedback designed to raise awareness of the visible stress responses that students display.

> Outside of physical things, looking at . . . operational tasks, include the human performance folks—so it could be some just repetitive task, it could be something that is pretty menial, but breaking it down to its essence. But then also adding stress to see . . . you know, continually meeting the standards while increasing the stress on this very specific aspect. (Leadership PJ/CRO Apprentice)

> But I think with that if the body is physically breaking down, it's putting that stress on the mind. So, you can stress out the mind, but maybe the body is not being stressed as much. But I think when you're stressing the body, it really starts to affect the mind as well. And you kind of get a twofold thing right there. (Leadership PJ/CRO Apprentice)

> So, I think just good feedback mechanisms [are needed], 360-degree feedback, opportunities for the team to do it internally. Which like I said, we are doing to an extent already, but I think that's probably the best way to encourage emotional regulation internal to the team environment. (Leadership PJ/CRO Apprentice)

How Students Are Doing on Stress Tolerance

Leadership and cadre indicated being aware when students were feeling stressed. They know what events are stressful and are familiar with the behavioral indicators displayed by students.

> These are things that would definitely stand out. If a student under stress showed the eye-rolling, head down, shouting, that's something that would definitely be annotated as a performance deficit. (Leadership PJ/CRO Apprentice)

> Yeah, I see that throughout the whole course. I think every course keeps them under some form of stress. It's kind of difficult for these guys to have a bad day without it showing. (Leadership TACP Apprentice)

While it does not appear to be a pervasive issue, leadership noted concerns with some students' ability to perform under stress without assistance and suggested that it would be helpful to build up Stress Tolerance prior to the course.

> Maybe the biggest one I would add is just operating in adversity. So, what I mean by that is, when things are not ideal, the kids can't operate. If they're getting yelled at, they can't operate. If the situation is not perfect, if a wrench gets thrown into it, they can't shift gears. (Leadership TACP ICC Phase II)

Teamwork

Across the courses, the goal is to develop the necessary individual skills and be prepared to apply them as a member of an operational team. Teamwork considerations include finding an appropriate balance of team-based and individual tasks during training, cultivating a sense of commitment to the team, and learning how to assess team contributions in individuals who may be inclined to stay quiet and follow along.

Participant comments suggest that Teamwork may be emphasized differently for certain career fields. For example, leadership from the PJ/CRO Apprentice course indicated that Teamwork is important, but they expect the trainees to be proficient in working as a team when they enter the course.

> And to be honest, this schoolhouse has evolved—from where it was 20 years ago. . . . The guys already came here with teamwork instilled, but we treated them as if it didn't exist. We became smarter. Like, hey man, we've got way too much information to give these guys to go over that area again, to keep wasting our time doing that if they've already got it. (Leadership PJ/CRO Apprentice)

> So, if it's not being instilled prior to, we're not deliberately doing it here. (Leadership PJ/CRO Apprentice)

> I don't think we emphasize it that much. Probably more on the monitoring and then . . . create a supportive environment that expresses interest in other's ideas and feedback. (Leadership PJ/CRO Apprentice)

Leadership from the TACP ICC clarified why it is essential for TACPs to understand how to integrate into an operational unit when deployed. During training, they are surrounded by TACPs; however, once in the field, they will be surrounded by operators from other services and specializations and will need to be able to integrate their capabilities in that context.

> The one thing from day 1, all the way through, that we drive is Teamwork. (Leadership TACP Apprentice)

> It's extremely—it's important, to his point about not being on a team of TACPs. If you're a TACP at any echelon, you are not a gray man anymore. (Leadership TACP ICC Phase I)

Opportunities for developing Teamwork among trainees were identified during the interviews. Some comments discussed a need to improve Teamwork among trainees, pushing the concept that individual behaviors affect team performance. Finding an optimal balance for using consequences was suggested to promote team cohesion and reinforce norms.

> Like I said, I think we're producing the healthiest and strongest folks we've produced. I just think we left behind our teamwork piece of that somehow—it's also cultural fear of people getting in trouble. (Leadership CCT STO SR Apprentice)

> So, I would say a simplified process that is still deliberate, more theater than anything, but just to reinforce that individual actions have team consequences. (Leadership PJ/CRO Apprentice)

How Students Are Doing on Teamwork

While not flagged as a pervasive issue, it was noted that the shift to an A&S versus historical approaches to training (i.e., Indoctrination course) has contributed to more emphasis being put on individual performance versus team performance.

> So, A&S guys versus Indoc guys—and this is PJs and CROs alike—definitely, there's a lack of teamwork. That's by far the most obvious characteristic. They're selected as individuals.

Trainability

In general, comments indicated that Trainability is evaluated throughout the various courses but is not deliberately developed. Participants indicated that they value individuals who can learn and apply information quickly. This attribute is recognized as being very important to success in the course, and it is expected that students will be trainable.

> Yeah, this one is a little bit tricky because we kind of assume that they come in with the skills to learn. (Leadership PJ/CRO Apprentice)

> So, I mean, we teach a hell of a lot, and we expect them to acquire it, but we aren't teaching them how to learn. That meta aspect of it is not there. (Leadership PJ/CRO Apprentice)

> I don't know that we're assessing whether or not they are capable of acquiring knowledge or if they are good at learning new skills. (Leadership TACP Apprentice)

It was suggested that trainability could potentially be improved by leveraging human performance (HP) experts during the pipeline.

> And actually, for this one, I would say HP integration again is a way to improve trainability. (Leadership PJ/CRO Apprentice)

How Students Are Doing on Trainability

In general, students are perceived as reasonably trainable, with some exceptions. Participants provided some examples of what causes students to struggle in a course. Student trainability was perceived to be affected by cognitive ability as well as the ability to take criticism and move on.

> And that's really the biggest issue that we've had is when you get somebody that just learns slow, there is too much information coming in to be a slow learner. There is not enough time in the day to make up for it with work ethic. You have to be smart; you have to be able to learn fast, and if you can't, you're going to fall behind. (Leadership PJ/CRO Apprentice)

> But what I mean by that, also, is just some of these kids just shut down if they get criticized. And I don't know if you would call that [needing to have a] thick skin . . . but yeah, I would just put that all under the umbrella of being able to do the job in less-than-ideal circumstances, being able to have thick skin, that sort of stuff. (Leadership TACP ICC Phase II)

Some Attributes Deliberately but Unevenly Developed

The attributes of Integrity, Communication, Physical Fitness, and Problem Solving are actively developed during the training courses. Interview participants described ways that each of these attributes are developed during training. The extent to which these attributes are intentionally developed depends on the course, career field, as well as whether the trainee is an officer or enlisted member.

Communication

Communication was not as frequently discussed relative to other attributes, but it is clearly valued across the career fields and is actively taught to some extent. Participant comments revealed some nuance regarding the types of context-specific communication perceived as important. For example, some comments indicated that this attribute is important for officers so that they can provide clear and succinct briefs regarding missions.

> We harp on them a lot, specifically in CRO Ops with their mission planning. One of the advantages of that two-week block before the actual course starts is we get to see them before their PJs get to see them for the most part. So, we can coach them through some leadership aspects. We make sure that they cover good, initial expectations with their team and help them through that. But we are coaching them on briefing style significantly during all of their mission planning briefs. (Leadership PJ/CRO Apprentice)

Technical and radio-based communications, described as being more procedural and technical in nature, was identified as being important for the TACP role specifically.

> I think we do this—I think this flows parallel with teamwork. Those are the two big things—like that is our—our primary job is communication. And our cadre know that. So . . . the first two things we hit is proper communication amongst your team and to instructors, and the teamwork within your team to make sure you're

> successful. And if those things are failing, we do correct those behaviors. (Leadership TACP Apprentice)

> I mean, overall, we absolutely have to do this to a great extent. We have to be . . . like, great communicators, both in execution and planning. But I don't know if we give formal training on, like, okay, active listening. (Leadership TACP ICC Phase I)

> During the CAS [Close Air Support] mission, they get feedback, like, hey, you told me this, but think about how that sounded. (Leadership TACP FTU Phase I)

However, Communication training on general skills, outside of mission-specific topics, is limited primarily to feedback and general classroom participation.

> We don't ask the enlisted guys to synthesize information and deliver it very often. Now, active listening, a traditional podium-style classroom environment, yes, they must take notes. Yes, they must be actively a part of the class, and they also have to demonstrate [they can] retain knowledge via a test. And other opportunities, this may be tough with a large class, but a student-delivered presentation based [on] material they received. (Leadership PJ/CRO Apprentice)

> When it comes to, like, interpersonal communication, like conveying points or building, like, an argument on something, I don't think we train that really at all. (Leadership TACP Apprentice)

> I wouldn't say it emphasizes [Communication], especially not any more than what they'd previously had [before entering this course]. (Leadership CCT STO SR STTS)

The role of Communication was also discussed in the context of team-based Problem Solving.

> We try to give them obvious answers because they're still learning it, but there's a couple different ways to take care of [developing Communication]. So that's a communication thing, but the application of it relies upon problem-solving skills. (Leadership PJ/CRO Apprentice)

Physical Fitness

Along with other attributes, students are expected to possess a level of fitness appropriate for the pipeline and their career field. Requirements for Physical Fitness training are generally integrated into the structure of training courses, supported by cadre and coaches, and formally evaluated using the Operator Fitness Test. Comments from interview participants also suggest that Physical Fitness training has improved over the years.

> I think this is the one area where we can say, yes, absolutely, we formally train and develop this. (Leadership TACP ICC Phase I)

> This is really important because once upon a time, like he was saying, we didn't have [this type of training]—we didn't have this growing up, so we worked out as we saw fit and we did the right thing; we tore our bodies down. And, of course, we couldn't perform for that long because we were always hurt. Now we have people . . . so functional movement is a huge one. What actually makes sense for your application of something else? So, the structure I think is huge, and I think we're definitely moving in the right direction, and it's pretty good overall here. (Leadership TACP ICC Phase I)

> I mean, physical fitness, I think by this time, it's almost looked at as an individual responsibility, which is similar to what you would see on an operational team. (Leadership CCT STO SR STTS)

> There's one physical fitness standard here. It's the Operator Fitness Test. That being said, none of our training evaluations are that physically demanding, where we start seeing people fail at that. So really, your annual Operator Fitness Test, you pass that, and then you PT in the morning with your team. (Leadership CCT STO SR STTS)

Some opportunities for improving Physical Fitness training across the courses may include ensuring the training is consistent throughout and emphasizing personal responsibility to train.

> So, I think it kind of goes up and down. It's not a steady. (Leadership TACP Apprentice)

> Because I know that you guys talk to more of them, like, you could physically develop very well with the coach's plan. I can also do the bare minimum, not develop at all, and then fail things. And that comes down to the individual. (Leadership TACP Apprentice)

Additionally, comments suggest that workouts could be more effectively designed and tailored to better suit the needs of individuals. Specialists with expertise are also considered critical to help address specific needs.

> And what they're running into now is logistical issues, because they have a stupid ass plan where they're going to have everybody doing deadlifts at the same time, and they have 45 guys. . . . There's probably 15 different ways to solve that problem. And now they're like, "Hey, we need a $10 million facility." I'm like, really? Really, that's what you need? Or do you need a half decent plan that people like to do? (Leadership PJ/CRO Apprentice)

> And in terms of developing them, again, I think this is really a core area for HP involvement to help develop. Because they have the expertise with PTs, with strength coaches and everything, just to help the students develop the strength and cardio that they need, as well as figuring out how we apply those to actual mission tasks. (Leadership PJ/CRO Apprentice)

How Students Are Doing on Physical Fitness

Students are expected to maintain appropriate fitness levels, and opportunities are built into each of the courses. In general, as an attribute, Physical Fitness was not perceived as an issue, and as the comment below demonstrates, it is not the only attribute needed to succeed.

> I think physically, with the human performance focus, physically we're getting stronger, fitter, healthier people, definitely, in my view. Now, the problem with that is that the human performance approach versus the beatdown approach, for lack of a better word, is that it doesn't build the mental resiliency and toughness that the old approach built with the teamwork. (Leadership CCT-STO-SR Apprentice)

Problem Solving

Comments indicate that developing and practicing Problem Solving is common throughout each course. In practice, Problem Solving often involves students working through mission-specific problems individually or as a team and determining the best course of action. Depending on the activity, the framing of Problem Solving in the event may be implicit in some cases or explicit in others.

> Yeah, I look at everything that a student is asked to do here is some level of problem solving. They may be fed the conditions for success, maybe, but how they get there . . . I mean, one, the troop-leading procedures and just aside from the actual receipt of mission, which may be handled up at the CRO level, I can see any of these guys working through all of these, from gathering facts, identifying key elements of the problem, offering solutions, gathering more intelligence. (Leadership PJ/CRO Apprentice)

> They do take contact possibly on the way there, or not, so they have to plan [for the possibility that] if they're going to get an ambush, how to react to ambush, and all that stuff in the cadre, [and] leave it all up to the team itself to figure out how they're going to do that. (Leadership TACP Apprentice)

> There was always a problem-solving aspect in this course throughout. I think the problems just get more difficult. (Leadership TACP FTU Phase I)

Some students, particularly officers or those in leadership roles within their team, may have more opportunities for higher-level Problem Solving.

> That's what I wrote: experiences may differ depending on what position you're in. You may get a lot of problem solving or get . . . all I did in tech school—I followed the guy in front of me until we got there. (Leadership TACP Apprentice)

Opportunities mentioned for development of Problem Solving included standardizing the training on steps to take when solving problems and giving students specific guidance on how to approach specific problems. Additionally, ensuring consistency and access to instructors who are adept at incorporating Problem Solving into the courses was mentioned as being potentially helpful. Skilled instructors should be able to find an optimal balance between the level of guidance that they provide and the freedom students have to exercise their skills.

> But I know on the officer side through a commissioning source you get training on problem solving. And then, like in a lot of NCO [non-commissioned officer] courses you'll get similar training. But it is a deliberate process for solving problems, so it is a no kidding, I'm going through a checklist in my mind or on paper or however they're doing it. (Leadership TACP Apprentice)

> Probably the one thing I would say that maybe can help develop those a little bit is just instructors stepping back and allowing them to tackle things themselves. (Leadership PJ/CRO Apprentice)

> So, yeah, it's subjective to the person doing it. Sometimes it's really great; they're with that instructor for an hour and a half to two hours. And sometimes they're with that instructor for 30 minutes. So, looking in, a lot of that is on us. (Leadership TACP ICC Phase I)

Summary of Leader, Instructor, and Student Perspectives

Some attributes such as Problem Solving and Communication are deliberately targeted for training during specific parts of the course. However, many other attributes such as Integrity and Trainability are expected to be present before students enter the training pipeline. Although some instructors identified opportunities to further develop these attributes, there was also a general belief that some attributes are more stable and would be difficult to change. Furthermore, instructors could readily observe behaviors indicating students were lacking in one or more of these attributes. A few areas of concern related to Stress Tolerance and Trainability. These attributes were also highlighted in performance evaluations, which we discuss in the next section.

Performance Evaluations

As part of a separate project evaluating the effectiveness of AFSPECWAR A&S, we collected instructor and peer evaluations of students in different training courses.[23] Table 3.4 provides a summary of the number of students evaluated by their peers and instructors.

Table 3.4. Number of Performance Evaluations Collected by AFSC

AFSC	Number Evaluated by Instructors	Number Evaluated by Peers
CCT/STO/SR	77	90
PJ/CRO	58	112
TACP/TACP-O	184	260

To better understand the reasons for these evaluations, we reviewed rater comments that were provided after responding to this statement: "Knowing what I know now, I would not want this person on my team." Comments were sometimes brief and not descriptive enough (e.g., wild card) to assign to a particular AFSPECWAR attribute. However, there was sufficient detail to assign the majority (approximately 85 percent) of comments to one or more attributes.

Most comments provided by instructors and peers indicating "I would not want this person on my team" were associated with four attributes: Teamwork, Drive, Trainability, and Stress Tolerance. In Figure 3.3, we present the total number of comments per attribute and provide example comments.

[23] See Krueger et al., 2024, for additional details.

Figure 3.3. Most Common Attributes (in Red) Associated with Poor-Performing AFSPECWAR Students

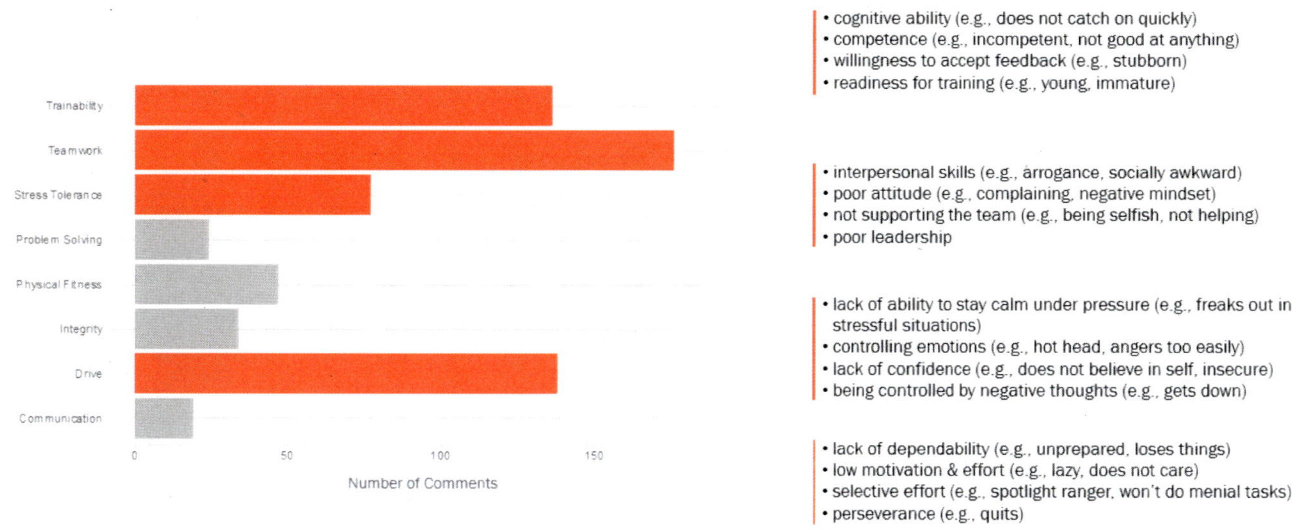

NOTE: Example comments for the attributes presented in gray are as follows:

- Problem Solving: Jumps to conclusions at times (PJ/CRO); unable to think critically and develop a plan to solve problems (TACP/TACP-O [Tactical Air Control Party Officer]).
- Physical Fitness: Not physically fit at this point for the job (CCT/STO/SR); does not have the physical ability (TACP/TACP-O).
- Integrity: Will lie, cheat, and steal (CCT/STO/SR); is a liar and cannot be trusted (PJ/CRO).
- Communication: Communication skills are weak (CCT/STO/SR); lacks basic communication skills (TACP/TACP-O).

Summary of AFSPECWAR Attribute Training Needs Analysis

To identify potential nontechnical training needs, we considered information from multiple sources, including quantitative performance ratings, raters' comments, and discussions with training instructors and select operational units. Although the emphasis varied by source, four attributes should be considered for further development. The four attribute gaps are Drive, Teamwork, Trainability, and Stress Tolerance. In many cases, deficiencies in one attribute may affect performance in a separate attribute. For example, a trainee who is considered lazy may lack the Drive necessary to perform individual tasks but may also lack the initiative to help teammates in need. Similarly, a trainee who has not learned the core job skills very well may also appear to perform poorly under a range of stressful conditions. Consequently, interventions designed to address one attribute may prove beneficial in addressing deficiencies in other attributes. In Chapters 4 through 7, we address strategies and evidence for developing these attributes to address current gaps. Because the gaps and training needs may change in the future, we also review research for developing the other nontechnical attributes that are currently not identified as gaps for AFSPECWAR.

Chapter 4

Developing Cognitive Attributes

Cognitive attributes are skills and abilities that are needed to learn (content and skills) and solve problems ranging in complexity. There are two AFSPECWAR cognitive attributes: Trainability and Problem Solving. In this chapter, we review concepts related to these attributes and review best practices for developing these attributes.

Trainability

Attribute Definition and Background

Table 4.1 presents the facets and definitions of Trainability.

Table 4.1. Trainability Facets and Definitions

Trainability	Demonstrates Capability to Learn and Execute Core Tasks and Duties
Learning Content	Acquiring knowledge (i.e., what you know)
Learning Procedures	Acquiring skills to execute and apply knowledge (i.e., what you do)

Of all the AFSPECWAR attributes, Trainability is the broadest and encompasses a wide range of constructs. Key factors influencing trainability include general cognitive ability, other traits, and training or instructional methods. It is important to note that Trainability is heavily screened by the USAF and AFPECWAR. There are minimum cognitive aptitude requirements to qualify for training, and then additional screening may occur during assessment and selection. Therefore, we focus primarily on training methods that further promote Trainability.

Concepts Related to Trainability

Performance Management

Research findings on the benefits of performance management (PM) have generally been disappointing. There appear to be many factors that impede organizational objectives to achieve desired outcomes from PM systems. Despite these challenges, several key lessons have been associated with better PM systems. These lessons include the following:[24]

Define the objective of the PM system (developmental, administrative).

- Encourage ratee participation in the PM process.

[24] Deidra J Schleicher., Heidi M. Baumann, David W. Sullivan, and Junhyok Yim, "Evaluating the Effectiveness of Performance Management: A 30-Year Integrative Conceptual Review," *Journal of Applied Psychology*, Vol. 104, No. 7, 2019, pp. 851–887.

- Provide training and easily accessible materials about the entire PM process (e.g., rating criteria, frequency of evaluations, outcomes of evaluations).
- Establish a system that ensures raters are unbiased and fair (e.g., rater training; behaviorally based, adequate opportunities to observe performance; objective and transparent criteria).
- Monitor and promote positive rater experiences (e.g., use formal processes that are accepted by raters), but more importantly, ensure that raters embrace the PM system (i.e., do it well).
- Promote development by emphasizing learning criteria rather than results.
- Maintain organizational commitment by increasing perceived fairness (e.g., due process).

Feedback

Providing feedback does not guarantee that feedback recipients will change their behavior.[25] In fact, feedback can lead to negative performance outcomes.[26] Nonetheless, feedback can be useful to direct attention, raise awareness of performance gaps, and provide information on how to address those gaps. Although some strategies such as sandwiching negative with positive feedback are appealing because of their simplicity, the effectiveness of feedback is more likely to be affected by the broader feedback environment, including the source (e.g., amount of expertise, trustworthiness), how the message is delivered (e.g., nonjudgmental), follow-on support and discussions with feedback providers, how much feedback-seeking is encouraged, and the quality of the message (accurate, informative).[27] Feedback delivery is difficult, but an important rule of thumb is to avoid statements about personality (e.g., is lazy) and focus on specific behaviors (e.g., clean equipment before storing).

Feedback Orientation

To better understand why some individuals respond well to feedback and others do not, we need to consider individuals' receptivity to feedback (i.e., feedback orientation) across four dimensions.[28] These four dimensions include perceived utility (e.g., "I find that feedback is critical for reaching my goals"), social awareness (e.g., "Using feedback, I am more aware of what people think of me"), feedback self-efficacy (e.g., "I know that I can handle the feedback that I receive"), and accountability (e.g., "It is my responsibility to apply feedback to improve my performance"). Although it may be possible to teach individuals how to receive feedback, there is very limited research in this area compared with strategies for delivering effective feedback.[29] However, there is ample research to suggest that providing frequent feedback combined with knowledge of performance standards is

---

[25] James W. Smither, Manuel London, and Richard R. Reilly. "Does Performance Improve Following Multisource Feedback? A Theoretical Model, Meta-Analysis, and Review of Empirical Findings." *Personnel Psychology*, Vol. 58, No. 1, 2005, pp. 33–66.

[26] Avraham N. Kluger and Angelo DeNisi, "The Effects of Feedback Interventions on Performance: A Historical Review: A Meta-Analysis, and a Preliminary Feedback Intervention Theory," *Psychological Bulletin*, Vol. 119, No. 2, 1996, pp. 254–284.

[27] Ian M. Katz, Rachel S. Rauvola, and Cort W. Rudolph, "Feedback Environment: A Meta-Analysis," *International Journal of Selection and Assessment*, No. 29, Vol. 3–4, 2021, pp. 305–325.

[28] M. London and J. W. Smither, "Feedback Orientation, Feedback Culture, and the Longitudinal Performance Management Process," *Human Resource Management Review*, Vol. 12, No. 1, 2002, pp. 81–100; Beth A. Linderbaum and Paul E. Levy, "The Development and Validation of the Feedback Orientation Scale (FOS)," *Journal of Management*, Vol. 36, No. 6, 2010, pp. 1372–1405.

[29] Rachel J. Ehrlich, Melissa R. Nosik, James E. Carr, and Byron Wine, "Teaching Employees How to Receive Feedback: A Preliminary Investigation." *Journal of Organizational Behavior Management*, Vol. 40, No. 1–2, 2020, pp. 19–29.

important to how individuals will respond to feedback.[30] That is, frequent feedback is more effective when individuals are aware of and understand the performance standards. This principle is already applied for technical training but may require additional planning and discussion among course instructors to implement for nontechnical attributes such as Teamwork and Communication.

Evidence-Based Practices for Developing Trainability

Three broad types of strategies have been developed to promote learning: (1) cognitive, (2) metacognitive, and (3) affective.[31] Cognitive strategies or tactics are approaches such as taking notes, summarizing, use of mental imagery, use of mnemonic devices, rehearsal, and other study skills that directly affect acquisition of content and procedural knowledge. Metacognitive strategies promote self-awareness and understanding about how one learns and solves problems.[32] There is evidence that metacognition can be taught by instructing individuals to engage in self-regulatory processes to monitor their progress, identify and correct errors, and think about what worked, under what conditions, and where improvements are needed.[33] Affective learning strategies focus on noncognitive aspects of learning, such as learner motivation, self-efficacy, and attributions for success and failure.[34] Affective learning comes about through interventions that foster students' beliefs about their abilities to learn, the value of studying the subject matter, or promoting internal attributions for success and external attributions for failure. There is much more research on cognitive and metacognitive than affective training interventions.

The effectiveness of training interventions for knowledge acquisition is typically evaluated by examining how individuals perform on a task sometime after participating in the intervention. Interventions are more effective when there is greater similarity between the training task and the outcome measure or testing task (e.g., learning the same list of words), which Hattie et al. refer to as reproductive outcomes.[35] For example, narrowly focused training interventions that provide instruction on a specific skill or technique, such as underlining or using a mnemonic device, have strong effects on performance outcomes. However, we expect that AFSPECWAR expects trainees to acquire transformational knowledge or deeper learning[36] (i.e., knowledge of content at a higher cognitive level that trainees can transfer or apply to other contexts). Moreover, we expect that AFSPECWAR's interest in facilitating deeper learning pertains to transfer of training to the job.

---

[30] Shaun Pichler, Gerard Beenen, and Stephen Wood, "Feedback Frequency and Appraisal Reactions: A Meta-Analytic Test of Moderators," *International Journal of Human Resource Management*, Vol. 31, No. 17, 2020, pp. 2238–2263.

[31] John Hattie, John Biggs, and Nola Purdie, "Effects of Learning Skills Interventions on Student Learning: A Meta-Analysis," *Review of Educational Research*, Vol. 66, No. 2, 1996, pp. 99–136.

[32] Janet E. Davidson, Rebecca Deuser, and Robert J. Sternberg, "The Role of Metacognition in Problem Solving," in J. Metcalfe and A. P. Shimamura, eds., *Metacognition: Knowing About Knowing*, MIT Press, 1994, pp. 207–226; John H. Flavell, "Metacognition and Cognitive Monitoring: A New Area of Cognitive-Developmental Inquiry," *American Psychologist*, Vol. 34, No. 10, 1979.

[33] Hattie, Biggs, and Purdie, 1996; National Research Council, *Education for Life and Work: Developing Transferable Knowledge and Skills in the 21st Century*, Washington, D.C.: National Academies Press, 2012.

[34] Hattie, Biggs, and Purdie, 1996.

[35] Hattie, Biggs, and Purdie, 1996.

[36] On transformational learning, see Hattie, Biggs, and Purdie, 1996; on deeper learning, see National Research Council, 2012.

There are several critical findings from research regarding cognitive, metacognitive, and affective learning. One key result is that learning depends on the content. For example, general training on how to use study skills typically shows small effects on cognitive outcomes; instead, such training is effective when it is taught in connection to specific subject matter.[37] Similarly, teaching metacognitive skills is most effective in the context of instruction on specific content, because the ability to assess one's performance is related to knowledge of the content area.[38]

A related finding is that transfer of learning depends on the context. The National Research Council (NRC) notes that characteristics of the job environment are the most important factor for transfer of skills learned in training to the job. Key aspects of the job environment to facilitate transfer include prompts or requirements to use skills learned in training on the job; opportunities for practice; and use of performance goals, incentives, and feedback.[39] Other important factors are a short delay between training and opportunities to apply learned skills on the job and follow-up after training.[40]

Another key finding pertains to the interdependencies among types of learning. For example, according to the NRC, training students on study tactics is more effective not only when it is taught with respect to specific content but when it incorporates metacognitive elements. Examples of metacognitive strategies include having trainees think out loud as they perform tasks, explain material to themselves, summarize material, justify their answers, and test themselves.[41] Affective learning is also associated with cognitive learning, and both types of learning can affect each other. For example, interventions that focus on affective learning promote cognitive learning, and interventions that foster cognitive learning can affect students' self-efficacy and attitudes toward studying and learning specific subject matter.[42]

There are also individual characteristics that affect trainability. For example, Hattie et al. found that overall, training interventions that focused on study skills were less effective for performance among adult learners (college-age and higher) than for younger students.[43] They conclude that older students who have developed study skills over many years are more resistant to change; however, training interventions have strong and positive effects on affective training outcomes, such as attitudes toward study, for adult learners.[44] In addition, metacognitive training interventions used in connection to specific content were effective for all outcomes (performance, study skills, and affect) regardless of the age of the trainees and were especially effective for older students. Conscientiousness is another individual characteristic that affects learning outcomes, with a positive correlation of conscientiousness

[37] Hattie, Biggs, and Purdie, 1996; National Research Council, 2012.

[38] National Research Council, 2012.

[39] Rebecca Grossman and Eduardo Salas, "The Transfer of Training: What Really Matters," *International Journal of Training and Development*, Vol. 15, No. 2, 2011, pp. 103–120; cf. National Research Council, 2012.

[40] National Research Council, 2012.

[41] National Research Council, 2012.

[42] Hattie, Biggs, and Purdie, 1996; National Research Council, 2012.

[43] Hattie, Biggs, and Purdie, 1996.

[44] Hattie, Biggs, and Purdie, 1996.

and grades at all ages, including college.[45] Poropat found that the strength of the association of conscientiousness and grades was similar to that of general cognitive ability and grades.[46]

In addition to the use of metacognitive approaches in combination with cognitive strategies, the NRC reviews other findings regarding effective instructional methods to facilitate deeper cognitive learning:[47]

- Prime or activate the motivation to learn. Examples include fostering "mastery" rather than "performance" goals (focusing goals on learning the material rather than performing well or avoiding poor performance), setting expectations that trainees will succeed and can learn the material, and drawing connections between the material and topics of interest to trainees
- Use problem-based learning methods rather than traditional lecture-based instruction[48]
- Use multimedia (i.e., presenting information using text and graphics rather than text alone)
- Provide opportunities for extensive practice with increases levels of difficulty
- Provide informative, explanatory feedback (not just feedback about whether responses are correct or incorrect).

Conclusions

Based on our review, training methods play a key role in trainability, and people can be trained on all three types of learning (cognitive, metacognitive, and affective). Key conclusions about these strategies to promote deeper learning include the following:

- Providing general instruction in cognitive, metacognitive, and affective strategies is generally not effective; instead, these strategies should be used in conjunction with specific content and contexts.
- Cognitive, metacognitive, and affective training strategies are generally not independent; instead, these strategies affect each other and should be used in concert.
- Characteristics of the job environment, particularly the presence of opportunities to apply and practice skills learned in training, along with clear goals, incentives, and feedback, are critical to promoting the transfer of training.

Problem Solving

Attribute Definition and Background

Table 4.2 presents the facets and definitions of Problem Solving.

---

[45] Arthur E. Poropat, "A Meta-Analysis of the Five-Factor Model of Personality and Academic Performance," *Psychological Bulletin*, Vol. 135, No. 2, 2009, pp. 322–338; National Research Council, 2012.

[46] Poropat, 2009.

[47] National Research Council, 2012. Note that NRC's review focuses primarily on K–12, with less emphasis on college students and other adults. Its review of meta-analyses and other research findings often does not distinguish whether results differ based on learner age.

[48] Some features of problem-based learning include use of small groups, self-directed learning, a focus on the problem as a way to achieve knowledge and skills, and use of facilitators or guides (see National Research Council, 2012).

Table 4.2. Problem Solving Facets and Definitions

Problem Solving	Identifying and Using Resources to Generate, Critically Evaluate, and Select Course(s) of Action
Problem Definition	• Describing the set of conditions that define task achievement • Gathering facts and insight from external or internal sources (e.g., personal experiences) • Identifying key elements of the problem (e.g., assumptions, constraints)
Solution Development	• Generating different, original, and potentially workable courses of action that may involve new perspectives (e.g., out-of-the-box thinking) • Predicting possible outcomes to identify flaws and revise courses of action • Developing possible future courses of action, prioritizing activities, and specifying tasks/subtasks
Solution Implementation	• Selecting a solution and, if needed, improving its workability • Acquiring and coordinating resources to implement a solution • Systematically checking progress toward task achievement

Given these facets, several concepts emerged in our literature review as key antecedents of Problem Solving, particularly relevant to the AFSPECWAR community. Table 4.3 includes definitions of these concepts, which we discuss in more detail in subsequent sections.

Table 4.3. Definitions of Concepts Related to Problem Solving

Concept	Definition
Critical thinking	The skills used in "the kind of thinking involved in solving problems, formulating inferences, calculating likelihoods, and making decisions" and the inclination to use those skills when "faced with problems to solve."[a]
Cognitive load	In the context of Problem Solving, the amount and magnitude of "cognitive-processing capacity" used to consider and analyze different information and data when deciding how to address a problem; cognitive load generally increases with more stress (e.g., time constraints, complexity, etc.).[b]
Far transfer	The application of "knowledge and skills . . . in a context that is very different from the context in which the skill was learned."[c]
Combat dynamic intuition	The ability to "think strategically in dynamic situations based on nonlinear knowledge," given operational demands "to think and act fast in (overly) complex situations with a high uncertainty, and where the consequences of flawed decisions may be dramatic."[d]
Naturalistic decisionmaking	"The study of how people use their experience to make decisions in field and 'real world' settings."[e]

SOURCES:

[a] Dawit T. Tiruneh, An Verburgh, and Jan Elen, "Effectiveness of Critical Thinking Instruction in Higher Education: A Systematic Review of Intervention Studies," *Higher Education Studies*, Vol. 4, No. 1, 2014, p. 2.

[b] John Sweller, "Cognitive Load During Problem Solving: Effects on Learning," *Cognitive Science*, Vol. 12, 1988, pp. 257–285.

[c] Raymond A. Dixon and Ryan A. Brown, "Transfer of Learning: Connecting Concepts During Problem Solving," *Journal of Technology Education*, Vol. 24, No. 1, Fall 2012, p. 4.

[d] Bjørn Tallak Bakken and Martin Gilljam. "Dynamic Intuition in Military Command and Control: Why It Is Important, and How It Should Be Developed," *Cognition, Technology & Work*, Vol. 5, 2003, p. 197.

[e] Julie Gore, Rhona Flin, Seville Stanton, and B. L. William Wong, "Applications for Naturalistic Decision-Making," *Journal of Occupational and Organizational Psychology*, Vol. 88, 2015, pp. 223–230.

Critical Thinking

Critical thinking comprises a broad range of skills used to identify and address a given topic or problem along with the dispositions to effectively use those skills in different contexts. In reviewing this range of skills and dispositions, Tiruneh et al. observed that "there is now a consensus that critical thinking, as a broad concept, involves both skills and dispositions" and that dispositions include "truth-seeking, open mindedness, systematicity, analycity, maturity, inquisitiveness, and self-confidence."[49] Thus, when considering critical thinking in the context of developing Problem Solving, such training or other programs should focus on developing relevant skills and content-area knowledge, and also participants' inclination and confidence to use these skills in new contexts.

Cognitive Load

Programs to develop Problem Solving must consider the cognitive load placed on participants in the context of meeting learning outcomes during such programs. Sweller observed that "domain specific knowledge in the forms of schemas [mental structures of knowledge and the relationships between them] is the primary factor distinguishing experts from novices in problem-solving skill."[50] He also noted that training programs that focus too heavily on specific, structured problem-solving methods (e.g., means-ends analysis) at the expense of developing schema and knowledge can increase participants' cognitive load and thus decrease the overall effectiveness of the training. Building on more recent research, Kalyuga et al. observe that for participants with lower levels of content-area knowledge, beginning with general problem-solving strategies instead of specific methods can help reduce cognitive load at the outset by providing "learners with generalized frameworks for managing load at the following stages."[51] They argue for developing flexible Problem Solving, with training that accounts for participants' beginning levels of both content-area knowledge and problem-solving skills.[52]

Far Transfer

The application of knowledge and skills to new and novel contexts involves the concept of far transfer. While participants in a training program should—after enough practice—be able to apply what they have learned to similar contexts, participants often have more difficulty doing so when facing entirely different contexts or environments.[53] This holds important implications for Problem Solving development.[54] Problem Solving development should be flexible such that it is responsive to

[49] Dawit T. Tiruneh, An Verburgh, and Jan Elen, "Effectiveness of Critical Thinking Instruction in Higher Education: A Systematic Review of Intervention Studies," Higher Education Studies, Vol. 4, No. 1, 2014, pp. 1–17.

[50] John Sweller, "Cognitive Load During Problem Solving: Effects on Learning," *Cognitive Science*, Vol. 12, 1988, p. 257.

[51] Slava Kalyuga, Alexander Renkl, and Fred Paas, "Facilitating Flexible Problem Solving: A Cognitive Load Perspective," *Educational Psychology Review*, Vol. 22, 2010, pp. 175–186.

[52] Kalyuga, Renkl, and Paas, 2010.

[53] Raymond A. Dixon and Ryan A. Brown, "Transfer of Learning: Connecting Concepts During Problem Solving," *Journal of Technology Education*, Vol. 24, No. 1, Fall 2012, pp. 2–17.

[54] These implications are consistent with Kalyuga, Renkl, and Paas, 2010.

participants' prior knowledge and experience with Problem Solving.[55] Furthermore, to develop participants' inclination to act effectively in new settings, such development and training programs should have opportunities to practice applying knowledge and skills in new and novel settings while still remaining in the training environment.

Active Learning and Adaptive Performance

Active learning has been linked to adaptive transfer and combines elements of other training interventions. Specifically, active learning is defined by two common principles: (1) trainees are self-guided in selecting exercises or training activities that they believe will promote their learning, and (2) trainees are encouraged to experiment with tasks themselves to identify the principles and approaches that are most effective. Several interventions have been developed to promote active learning that emphasize specific training design elements, including stimulating metacognitive processes to encourage exploration, framing errors as a critical part of the learning process, and using internal (e.g., self-management) and external feedback (e.g., positive feedback) to manage trainees' emotions as they engage in the learning process.[56]

Combat Dynamic Intuition

Cognitive load is not only an important consideration for Problem Solving development, but it is also critical for participants' ability to grapple with high cognitive loads in complex, real-world contexts such as combat environments. Bakken and Gilljam discuss combat dynamic intuition as the ability to "think strategically in dynamic situations based on non-linear knowledge," alluding to far transfer of knowledge and skills to such "dynamic" and changing situations.[57] The high cognitive loads placed on those in such complex environments has important implications for Problem Solving development; training and development programs should consider the cognitive loads placed on participants during training, while also developing participants' inclination to use skills to effectively manage cognitive load in real-world settings.[58]

Naturalistic Decisionmaking

For ongoing research to improve programs that develop Problem Solving skills, training leaders should consider naturalistic decisionmaking, "the study of how people use their experience to make decisions in field and 'real world' settings" as opposed to strictly controlled laboratory settings.[59] With a focus on participants' expertise and prior experience and knowledge, naturalistic decisionmaking encompasses various methods to elicit expert knowledge and uncover how participants use prior

---

[55] Kalyuga, Renkl, and Paas, 2010.

[56] Bradford S. Bell and Steve W. J. Kozlowski, "Active Learning: Effects of Core Training Design Elements on Self-Regulatory Processes, Learning, and Adaptability," *Journal of Applied Psychology*, Vol. 93, No. 2, 2008, pp. 296–316.

[57] Bjørn Tallak Bakken and Martin Gilljam, "Dynamic Intuition in Military Command and Control: Why It Is Important, and How It Should Be Developed," *Cognition, Technology & Work*, Vol. 5, 2003, p. 197.

[58] Sweller, 1988; Kalyuga, Renkl, and Paas, 2010.

[59] Julie Gore, Rhona Flin, Seville Stanton, and B. L. William Wong, "Applications for Naturalistic Decision-Making," *Journal of Occupational and Organizational Psychology*, Vol. 88, 2015, p. 223.

knowledge and experience to inform their decisions and actions in field environments. For example, when developing a training program for the USAF Guardian Angel Weapon System (GAWS), Militello et al. applied several techniques from cognitive task analysis.[60] First, they became familiar with GAWS by attending several PJ workshops and trainings and by reviewing existing literature and training manuals to derive cognitive requirements for GAWS. Next, the authors elicited knowledge from PJ, CROs, and survival-evasion-resistance-escape specialists through two related interview techniques: one is Critical Decision, which allows for "identifying decision requirements in the context of real-world events"; the other is Knowledge Audit, which "includes question probes derived from the expertise literature intended to elicit specific examples of expertise" such as "examples of improvisation and key mental models required for proficient performance."[61] In discussing its applications, Gore et al. note that naturalistic decisionmaking can be helpful in exploring how individuals make decisions in complex environments with unclear opportunities, constraints, and stressors.[62] Thus, the methods used in naturalistic decisionmaking can be helpful to examine the effectiveness of Problem Solving training and inform continuous improvement.

Evidence-Based Practices for Developing Problem Solving

Cognitive Load Is a Critical Factor to Consider

In general, when designing a training program for participants with low or novice levels of content-area knowledge and Problem Solving skills with which to apply that knowledge, the evidence in the literature supports first focusing on development of content-area knowledge and expertise before shifting the focus to specific and structured Problem Solving skills; such a program could consider including general Problem Solving frameworks in the earlier stage to aid in bolstering content knowledge.[63] This chronological organization of training efforts over time generally corresponds with existing AFSPECWAR training design (see Figure 4.1).

As previously discussed, this chronological organization accounts for higher cognitive load—a critical moderating factor—that can detract from the effectiveness of an overall training program if too much emphasis is placed on specific Problem Solving skills before the requisite content-area knowledge. Instead, by first developing a strong foundation in content-area knowledge and associated schemata—

Figure 4.1. Notional Development of Content Knowledge and Problem Solving Skills over Time

	Problem Solving Skills	
	Low	High

[60] Laura G. Militello, Christen E. Sushereba, Matthieu Branlat, Robert Bean, and Victor Finomore, "Designing for Military Pararescue: Naturalistic Decision-Making Perspective, Methods, and Frameworks," *Journal of Occupational and Organizational Psychology*, Vol. 88, 2015, pp. 251–272.

[61] Militello et al., 2015, pp. 255–256.

[62] Gore et al., 2015.

[63] Sweller, 1988; Jim Jozwiak, "Teaching Problem-Solving Skills to Adults," *MPAEA Journal of Adult Education*, Vol. 33, No. 1, Spring 2004, pp. 19–34.

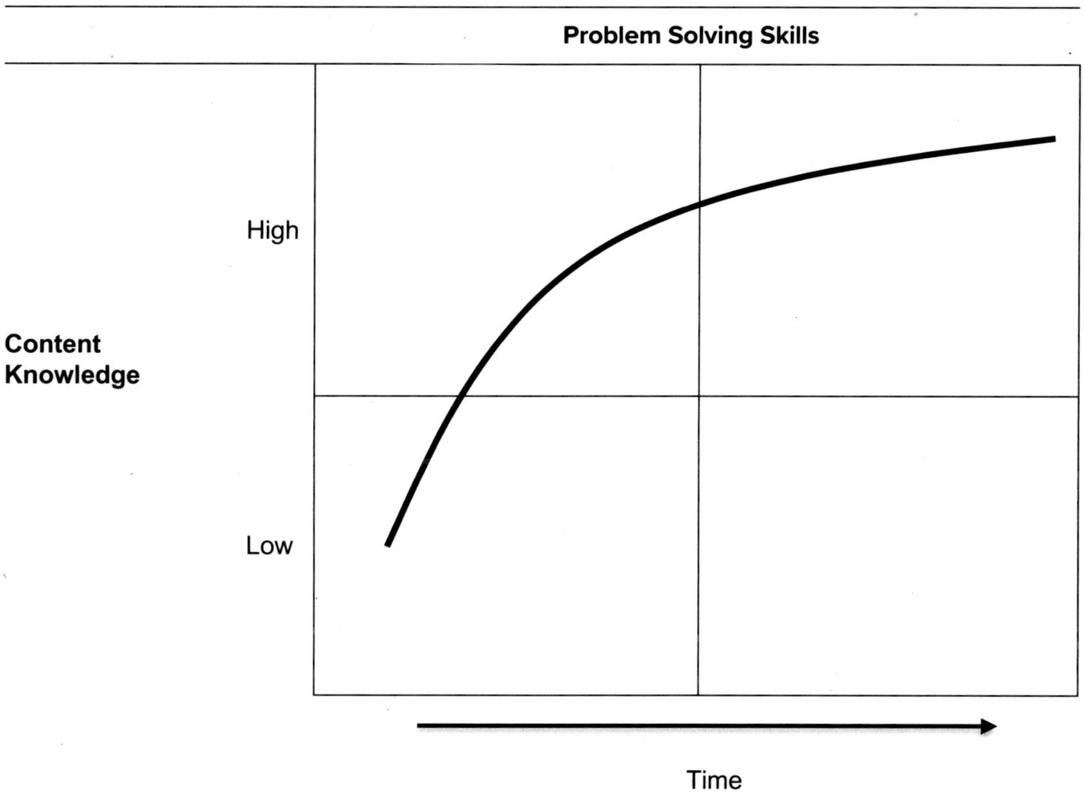

especially if such content areas are increasingly complex or technical—a training program can prepare participants for effective learning of specific Problem Solving skills and strategies later on.[64]

The following practices include evidence from experimental or quasi-experimental studies. Therefore, while these studies might have evidence for supporting a causal relationship between the training and participant outcomes, it is also important to consider the applicability of these research findings to real-world conditions.

Developing Core Knowledge with General Problem-Solving Frameworks as Instructional Tools

With novice learners, more recent research into knowledge transfer has emphasized the importance of focusing first on developing content knowledge with the use of general Problem Solving strategies instead of more complex or domain-specific strategies.[65] Recent reviews of the literature highlight how bolstering novice learners' content knowledge allows them to be better prepared for specialized

---

[64] Tanmay Sinha, and Manu Kapur, "When Problem Solving Followed by Instruction Works: Evidence for Productive Failure," *Review of Educational Research*, Vol. 91, No. 5, October 2021, pp. 761–798; Greg Ashman, Slava Kalyuga, and John Sweller, "Problem-Solving for Explicit Instruction: Which Should Go First When Element Interactivity Is High?" *Educational Psychology Review*, Vol. 32, 2020, pp. 229–247.

[65] This newer research has been consistent with Sweller, 1988; Kalyuga, Renkl, and Paas, 2010.

learning and application later on.[66] Another study describes NASA's transformation of its training program for International Space Station flight controllers who must have both high technical expertise and high proficiency with Problem Solving skills when facing novel situations.[67] Consistent with other findings in the literature, NASA decided to focus first on developing "basic technical knowledge and Space Flight Resource Management (SFRM) skills, such as decision making and teamwork" and later on "more advanced technical and SFRM skills" that would also be "practiced in discipline-specific simulations."[68]

Minimalist Decision Training

As opposed to traditional commander trainings with operational and dynamic factors in play, minimalist decision training holds "static" factors constant so that participants can focus on dynamic factors in a situational environment. The goals of this training method are to allow participants to focus on immediate needs in a mission context; allow for more opportunities for training by minimizing time spent on more traditional, longer strategy trainings; and ultimately develop participants' combat dynamic intuition. Reflecting on the results of an experiment testing the effectiveness of minimalist decision training, Bakken and Gilljam observed that study participants from the military academy and business schools often developed a routine for how they would approach successive problems; therefore, they suggested increasing the intensity and frequency of training and providing guidance to encourage participants to reflect on their decisionmaking process.[69]

"Think Like a Commander" Training

A complement to combat dynamic intuition, with implications for far transfer to new contexts, is adaptive thinking, which is defined in the military context as "the cognitive behavior of an officer who is confronted by unanticipated circumstances during the execution of a planned military operation."[70] One approach to improving adaptive thinking for service members is Think Like a Commander training, in which participants are presented with successive audiovisual vignettes or "cognitive battle drills" of a complex tactical scenario along with indicators for the participants to consider when

---

[66] Slava Kalyuga and John Sweller, "Cognitive Load and Expertise Reveal," in K. A. Ericsson, R. R. Hoffman, A. Kozbelt, and A. M. Williams, eds., *The Cambridge Handbook of Expertise and Expert Performance*, 2nd ed., Cambridge University Press, 2018, pp. 793–811; John Sweller, Jeroen J. G. van Merriënboer, and Fred Paas, "Cognitive Architecture and Instructional Design: 20 Years Later," *Educational Psychology Review*, Vol. 31, 2019, pp. 261–292.

[67] Lynne Martin, William O'Keefe, Lacey Schmidt, Immanuel Barshi, and Robert Mauro, "Houston, We Have a Problem Solving Model for Training," *Journal of Organizational Psychology*, Vol. 12, No. 1, 2012, pp. 57–68.

[68] Martin et al., 2012, p. 58.

[69] Bakken and Gilljam, 2003.

[70] James W. Lussier, Karol G. Ross, and Bob Mayes, "Coaching Techniques for Adaptive Thinking" (paper presented at the meeting of the Interservice/Industry Training, Simulation, and Education Conference, Orlando, Fla., December 2000); Scott B. Shadrick, James W. Lussier, and Christopher Fultz, *Accelerating the Development of Adaptive Performance: Validating the Think Like a Commander Training*, Arlington, Va.: U.S. Army Research Institute for the Behavioral and Social Sciences, ARI Research Report 1868, 2007.

making their decisions.[71] For each vignette, participants record a list of factors that should be considered before deciding on further action; furthermore, the time limit for listing factors is reduced for each vignette. Finally, an instructor leads a reflective discussion on what factors the participants listed, why they listed them, and how their lists compare with those of expert commanders. In one study with participants with and without experience in Operation Iraqi Freedom or Operation Enduring Freedom, researchers found that officers with experience in these operations "performed significantly better" in the training, thus suggesting that this training improves "a behavior that is relevant to the adaptive performance required by tactical leaders in current conflicts."[72] Another study, conducted with students in the Armor Captains Career Course, found that participants who received Think Like a Commander training performed significantly better at completing a company operations order than participants who did not receive the training.[73]

Consider the Opposite; Take an Outside View

In their review of experimental studies of strategies to improve decisionmaking, Dalal and Bolunmez note two additional practices that have been shown to support generation and consideration of new ideas: consider the opposite, in which participants purposefully brainstorm potential reasons why a decision might be incorrect; and take an outside view, in which participants look to comparable, prior decisions and use them to inform their current decision.[74] These practices can prompt participants to examine new ideas and information and thus prepare them to address a wider range of possibilities when making decisions.

Other Frameworks with Mixed Evidence

During our review, we encountered other frameworks that, though seemingly prevalent in the Problem Solving literature, are accompanied by mixed evidence on their effectiveness or usefulness. For future reference, we share them with a brief description of the framework and the caveats in the evidence base (Table 4.4). While elements of these frameworks might be useful in certain contexts, their use should be informed by these caveats and by future research into these frameworks.

---

[71] Shadrick, Lussier, and Fultz, 2007.

[72] Shadrick, Lussier, and Fultz, 2007, p. 1.

[73] Scott B. Shadrick, Brian T. Crabb, James W. Lussier, and Thomas J. Burke, *Positive Transfer of Adaptive Battlefield Thinking Skills*, Arlington, Va.: U.S. Army Research Institute for the Behavioral and Social Sciences, ARI Research Report 1873, 2007.

[74] Reeshad S. Dalal, and Balca Bolunmez, "Evidence-Based Strategies to Improve Workplace Decisions: Small Steps, Big Effects," Alexandria, Va: Society for Human Resources Management, 2016.

Table 4.4 Additional Problem Solving Frameworks

Framework	Description	Evidence
Problem-based learning[a]	Originated from medical education and focuses on student-centered and student-directed learning to define the problem, develop hypotheses and tests, and gather and evaluate data and evidence.	Savery also notes that at the time of writing, more rigorous empirical research had been mixed—a concern that continues to exist as highlighted by more recent meta-analyses.[b]
Perceptual-cognitive skills training[c]	Focuses on developing participants' ability to learn from a new environment, apply existing skills and knowledge, and decide on the best course of action.	In a more recent review of the literature on perceptual-cognitive skills training, Renshaw et al. note that the empirical evidence for far transfer is lacking; so is critique-related evidence for methodological rigor.[d]
Constraint management[e]	Focuses on developing skills to identify and creatively address constraints in a given context.	Medeiros et al. included an important caveat that while constraint management is grounded in leadership theory, their propositions on constraint management "should be interpreted with caution—more as an extension of theory that requires future research attention than as a model with substantive empirical support."[f]

SOURCES:
[a] John R. Savery, "Overview of Problem-Based Learning: Definitions and Distinctions," *Interdisciplinary Journal of Problem-Based Learning*, Vol. 1, No. 1, Spring 2006, pp. 9–20.
[b] Savery, 2006; Yongjie Qin, Yungui Wang, and Robert E. Floden, "The Effect of Problem-Based Learning on Improvement of the Medical Educational Environment: A Systematic Review and Meta-Analysis," *Medical Principles and Practice*, Vol. 25, August 2016, pp. 525–532; Philip Hallinger and Edwin M. Bridges, "A Systematic Review of Research on the Use of Problem-Based Learning in the Preparation and Development of School Leaders," *Educational Administration Quarterly*, Vol. 53, No. 2, 2017, pp. 255–288.; Yong Liu and Attila Pásztor, "Effects of Problem-Based Learning Instructional Intervention on Critical Thinking in Higher Education: A Meta-Analysis," *Thinking Skills and Creativity*, Vol. 45, Art. 101069, September 2022, pp. 1–21.
[c] Paul Ward, Damian Farrow, Kevin R. Harris, A. Mark Williams, David W. Eccles, and K. Anders Ericsson, "Training Perceptual-Cognitive Skills: Can Sport Psychology Research Inform Military Decision Training?" *Military Psychology*, Vol. 20, Suppl. 1, 2008, pp. S71–S102.
[d] Ian Renshaw, Keith Davids, Duarte Araújo, Ana Lucas, William M. Roberts, Daniel J. Newcombe, and Benjamin Franks, "Evaluating Weaknesses of 'Perceptual-Cognitive Training' and 'Brain Training' Methods in Sport: An Ecological Dynamics Critique," *Frontiers in Psychology*, Vol. 9, Art. 2468, January 2019, pp. 1–14.
[e] Kelsey E. Medeiros, Logan L. Watts, and Michael D. Mumford, "Thinking Inside the Box: Educating Leaders to Manage Constraints," in Chunfang Zhou, ed., *Handbook of Research on Creative Problem-Solving Skill Development in Higher Education*, Hershey, Pa.: IGI Global, 2017, p. 40.
[f] Medeiros, Watts, and Mumford, 2017.

Conclusions

The following sections present the key conclusions and takeaways from our review on Problem Solving for AFSPECWAR.

Focus First on Developing a Knowledge Base When Participants Have Beginner Levels of Content Knowledge

For a given training program, begin by understanding the trainees' level of content knowledge, especially in complex or technical subjects. For those with beginner levels of content knowledge, focus on developing their knowledge base first, including general Problem Solving strategies as part of the broader goal of developing content knowledge. Then, after participants become more proficient with their content knowledge, the training can shift focus, emphasizing specific Problem Solving strategies and providing opportunities to apply content knowledge to novel contexts and scenarios.

Be Mindful of Cognitive Load

When considering both the chronological order of training elements and the content of these training elements, the training should account for the cognitive load that participants are expected to face and whether it will help or hinder the training's effectiveness. In some instances, like stress-exposure training, increasing cognitive load through the application of stressors can be helpful in supporting participants' learning and ability to apply knowledge to different, novel contexts. Furthermore, later development of specific Problem Solving frameworks for complex or domain-specific environments can be useful to reduce participants' cognitive load during these periods of training—and eventually when facing these environments in the field.

Training Can Be Tailored to Unique Training and Operational Needs with an Emphasis on Continuous Improvement of Training Design

The different practices described in our review open various avenues to develop different training regimens based on situational and operational needs and with consideration for cognitive load and bolstering the likelihood of far transfer. In addition, while these practices are based on context-specific experimental studies, their underlying principles can be adapted to different training needs and possibly be used for making future modifications to training design and execution. The methods used in naturalistic decisionmaking can be helpful to inform improvements and changes to trainings (e.g., GAWS).[75] It is also important to note that other training techniques that promote adaptive expertise may need to be considered for operators with more experience, following the formal training pipeline.[76]

---

[75] Militello et al., 2015.

[76] Johnathan K. Nelson, Stephen J. Zaccaro, and Jeffrey L. Herman. "Strategic Information Provision and Experiential Variety as Tools for Developing Adaptive Leadership Skills," *Consulting Psychology Journal: Practice and Research*, Vol. 62, No. 2, 2010.

Chapter 5

Developing Physical Fitness

In this chapter, we review evidence-based practices for developing as well as maintaining physical fitness.

Attribute Definition and Background

Physical Fitness training can be categorized in multiple ways: ability (e.g., cardiorespiratory endurance, upper-body strength), movement pattern (e.g., running, lifting), desired goal (e.g., marathon training, task, functional training, injury prevention), equipment requirements (e.g., body weight, free weights, machines), and energy systems (e.g., aerobic, anaerobic). Considering the requirement that AFSPECWAR candidates must have a baseline level of fitness to complete assessment and selection (and training), we adopt a goal-oriented perspective to focus on training interventions designed to improve specific types of military movements (e.g., rucking), maintenance of fitness, and injury prevention. We considered these goals along with the broad definition previously provided for AFSPECWAR (see Table 5.1 for the facets and definitions of Physical Fitness).

Table 5.1. Physical Fitness Facets and Definitions

Physical Fitness	Develops and Maintains Physical Capability to Train and Execute Job Duties
Strength	• Demonstrating muscular power, tension, and endurance to apply or resist force (e.g., lifting, pushing)
Cardiorespiratory Endurance	• Demonstrating efficient delivery of oxygen to large muscle groups to sustain muscular activity over prolonged periods
Functional Movement	• Demonstrating the full range of motion, balance, and coordination needed to perform skilled movements

It is also important to note that AFSPECWAR has already allocated considerable resources to physical fitness, injury prevention, and rehabilitation. For example, AFSPECWAR has strength and conditioning coaches and physical therapists who provide expert-level guidance and programming. These resources are reflected in the heavy emphasis in the early preparation of AFSPECWAR candidates in SWCC and in follow-on requirements to pass the Candidate Fitness Test and the Operator Fitness Test.[77] Early training and periodic fitness testing suggest that students entering

[77] Krueger et al., 2022.

initial skills training will have a strong baseline from which to further develop their fitness and physical skills for specific job tasks.

Consequently, an important goal for many trainees may be to maintain the fitness they gained while focusing on learning how to perform their occupational specialty technical tasks. Therefore, we begin with an overview of evidence-based practices to maintain fitness, followed by discussions of interventions to develop task-specific fitness. We then conclude with a discussion of injury prevention strategies.

Evidence-Based Practices

Maintaining Physical Fitness

A review of research suggested several guidelines for preserving strength and endurance (Table 5.2).[78] It is important to note that the guidelines are primarily drawn from general population research rather than research specifically conducted with military personnel. It is possible that maintaining a higher level of fitness in the military and specifically within SW may require modifications to the frequency, volume, and intensity of training. Additionally, some findings were limited to one or just a few studies, which need to be confirmed in independent studies. Finally, the guidelines represent modifications in a single factor (e.g., reduced frequency or reduced volume, but not both). Considering these limitations, the intensity of workouts appears to be the most important factor. Therefore, if training frequency and volume need to be reduced because of student schedules, trainers should ensure that sessions include high-intensity workouts.

Table 5.2. Guidelines to Maintain Cardiorespiratory Endurance and Muscular Strength

Ability	Training Factor	Guideline
Endurance	Frequency	At least two training sessions per week for maintenance up to 15 weeks; in trained populations, reduce frequency by no more than 20 percent
	Volume	Long-term endurance (e.g., one to three hours) maintained when volume reduced by 33 percent
	Intensity	Maintain intensity of workouts; more important to maintain compared with frequency and volume
Strength	Frequency	At least one session per week to maintain one repetition maximum
	Volume	Limited research suggests at least one set per training session to maintain strength
	Intensity	Limited research suggests near maximal strength (or at least maintain intensity similar to typical training sessions)

[78] Barry A. Spiering, Iñigo Mujika, Marilyn A. Sharp, and Stephen A. Foulis, "Maintaining Physical Performance: The Minimal Dose of Exercise Needed to Preserve Endurance and Strength Over Time," *The Journal of Strength & Conditioning Research*, Vol. 35, No. 5, 2021, pp. 1449–1558.

Developing Physical Fitness

Other research has emphasized the importance of adding specificity to training because the type and level of physical demand varies across job tasks. For example, the physical abilities required to exfil on a rope ladder to a helicopter will be different from the abilities required to conduct a ground movement while carrying a heavy load (e.g., rucking). In addition to differences in task demands, there are also differences between individuals' fitness levels and adaptations that also require additional specificity of physical fitness programming. Knapick et al. conducted a systematic review of load carriage performance.[79] Building on that study, Vaara et al. conducted a review of interventions designed to improve physical performance on specific military tasks, which included load carriage (i.e., rucking) and casualty evacuation.[80] Their review suggests the following key findings:

- Performance gains are greatest when training combines whole-body strength and aerobic training.
- Improvements can be seen with three training sessions per week for four weeks.
- Although aerobic training improves performance, strength training is relatively more important.
- Higher-intensity and lower-volume training may be an effective training strategy, especially when compared with traditional approaches that emphasize longer periods of moderate-intensity training.[81]

More specific guidance has been provided as part of a separate review on load carriage training. The recommendations are as follows:

> - "two to four evenly spaced load carriage sessions per month;
> - carried loads that are initially light yet progress in weight to meet that required for given military tasks;
> - load carriage task durations and distances that gradually increase (yet not at the same time as increase in load) to meet military requirements
> - periods of recovery spaced throughout the program to allow the body to recover from the conditioning stimulus; and
> - supplemental conditioning (muscle strength and aerobic training) sessions utilizing functional movement patterns to provide adaptation to a broad spectrum of load carriage duties and tasks."[82]

---

[79] Joseph J. Knapik, Everett A. Harman, Ryan A. Steelman, and Bria S. Graham, "A Systematic Review of the Effects of Physical Training on Load Carriage Performance," *Journal of Strength and Conditioning Research*, Vol. 26, No. 2, 2012, pp. 585–597.

[80] Jani P. Vaara, Herbert Groeller, Jace Drain, Heikki Kyröläinen, Kai Pihlainen, Tommi Ojanen, Chris Connaboy, Matti Santtila, Philip Agostinelli, and Brad C. Nindl, "Physical Training Considerations for Optimizing Performance in Essential Military Tasks," *European Journal of Sport Science*, Vol. 22, No. 1, 2022, pp. 43–57.

[81] Simon D. Burley, Jace R. Drain, John A. Sampson, Bradley C. Nindl, and Herbert Groeller, "Effect of a Novel Low Volume, High Intensity Concurrent Training Regimen on Recruit Fitness and Resilience," *Journal of Science and Medicine in Sport*, Vol. 23, No. 10, pp. 2020979–2020984.

[82] Robin M. Orr, Rodney Pope, Venerina Johnston, and Julia Coyle, "Load Carriage: Minimising Soldier Injuries Through Physical Conditioning-A Narrative Review," *Journal of Military and Veterans Health*, Vol. 18, No. 3 2010, p. 37.

These guidelines have been generally supported in other reviews of military training. As summarized by Kyröläinen et al., "achieving optimal physiological adaptations and, therefore, more effective development in physical performance, increasing attention should be paid to progressive and individualized training programs and their division into phases that sequentially develop performance. Thus," the researchers note, "a personalized approach to performance optimization should be emphasized when improving physical fitness and enhanced operational readiness" among military personnel.[83]

Other quantitative reviews have focused on comparing traditional military approaches (e.g., push-ups, calisthenics) versus nontraditional approaches (e.g., resistance training such as deadlifts and squats) to fitness training. For example, a recent meta-analysis found nontraditional training was more effective for improving on some but not all abilities.[84] Specifically, nontraditional training yielded greater improvements in muscular endurance, power, and occupational performance. There were no differences in improvements between training type for aerobic fitness or flexibility. These findings may be less relevant to AFSPECWAR, which has additional resources and experts available to provide specialized programming compared with the general-purpose force.

Broader population meta-analyses have also supported the importance of resistance training for increasing muscle size, especially when compared directly against only aerobic training.[85] A separate systematic review found that combining aerobic training with resistance training did not impede gains in muscle size or strength; however, there was evidence to suggest that aerobic and resistance training should be completed in separate sessions if the goal of training was to develop explosive strength.[86]

Conclusions

These studies highlight the importance of establishing the purpose of training to guide selection of the most appropriate programming. For example, the specific exercises, number of repetitions, frequency, and intensity may need to be varied to achieve different fitness goals (e.g., maintenance versus development of muscular strength). Furthermore, training should be individualized and progressively focus on job-relevant physical demands.

---

[83] Heikki Kyröläinen, Kai Pihlainen, Jani P. Vaara, Tommi Ojanen, and Matti Santtila, "Optimising Training Adaptations and Performance in Military Environment," *Journal of Science and Medicine in Sport*, Vol. 21, No. 11 2018, p. 1137.

[84] Chelsea Smith, Kenji Doma, Brian Heilbronn, and Anthony Leicht, "Effect of Exercise Training Programs on Physical Fitness Domains in Military Personnel: A Systematic Review and Meta-Analysis," *Military Medicine*, No. 9/10, September 2022, pp. 1065–1073.

[85] Jozo Grgic, Luke C. Mcllvenna, Jackson J. Fyfe, Filip Sabol, David J. Bishop, Brad J. Schoenfeld, and Zeljko Pedisic, "Does Aerobic Training Promote the Same Skeletal Muscle Hypertrophy as Resistance Training? A Systematic Review and Meta-Analysis," *Sports Medicine*, Vol. 49, No. 2, 2019, pp. 233–254.

[86] Moritz Schumann, Joshua F. Feuerbacher, Marvin Sünkeler, Nils Freitag, Bent R. Rønnestad, Kenji Doma, and Tommy R. Lundberg, "Compatibility of Concurrent Aerobic and Strength Training for Skeletal Muscle Size and Function: An Updated Systematic Review and Meta-Analysis," *Sports Medicine*, November, Vol. 10, 2021, pp. 1–12.

Chapter 6

Developing Intrapersonal Attributes

Intrapersonal attributes are skills and abilities "that reside with the individual."[87] They include motivational constructs (i.e., Drive), morals and ethics (i.e., Integrity), and performance and emotion regulation under stress (i.e., Stress Tolerance). In this chapter, we review concepts related to these attributes and review best practices for developing them.

Drive

Attribute Definition and Background

Table 6.1 presents the facets and definitions of Drive.

Table 6.1. Drive Facets and Definitions

Drive	Initiates Action and Persists to Accomplish Tasks and Goals to High Standards
Initiative	• Taking action by devoting physical or mental resources to task accomplishment
Persistence	• Sustaining effort over time (i.e., hardworking, grit)
Reliability	• Paying attention to details, taking responsibility, and committing to performing high-quality work

These concepts can be considered antecedents or key predictors of Drive, which aligns with the concept of work effort or "how hard workers try to perform their jobs."[88] We briefly describe some of these concepts below.

[87] National Research Council (US) Committee on the Assessment of 21st Century Skills, "Assessing Intrapersonal Skills," in *Assessing 21st Century Skills: Summary of a Workshop*, Washington, D.C.: National Academies Press, 2011, p. 63.

[88] Chad H. Van Iddekinge, John D. Arnold, Herman Aguinis, Jonas W. B. Lang, and Filip Lievens, "Work Effort: A Conceptual and Meta-Analytic Review," *Journal of Management*, Vol. 49, No. 1, 2023, 125–157. Van Iddenkinge et al. treat *work effort* as a separate concept, one that motivation, conscientiousness, and other factors can predict. Much of the literature on work motivation, however, includes definitions and terms related to effort. For that reason, we combine the other key concepts under the header of motivation *and* effort.

Concepts Related to Drive

Conscientiousness

Conscientiousness is a personality characteristic that involves several facets, such as dependability and orderliness.[89] In general, highly conscientious people work hard to achieve their goals; they are organized and reliable. Conscientiousness is also one of the most consistent predictors of important work-related and career outcomes such as job performance, income, and occupational status.[90] Although personality characteristics like conscientiousness can change over one's lifetime, such changes tend to be small in adulthood.[91] Therefore, training interventions focused on personality characteristics like conscientiousness may not be all that effective.[92] However, personality assessments can be a useful tool to promote self-awareness; therefore, formal development programs often include them as part of a broader strategy to help individuals identify their strengths and weaknesses.[93]

Work Motivation and Effort

Although there is no single theory for work motivation, Mitchell and Daniels organize theories of work motivation into two separate groups: those that concentrate on internal processes (e.g., beliefs, emotions) and those that concentrate on external factors (e.g., work environment constraints).[94] Another important distinction in the work-motivation literature is between intrinsic and extrinsic motivation. Intrinsic motivation derives from the enjoyment or pleasure of engaging in activities for their own sake, whereas extrinsic motivation entails engaging in activities to achieve specific outcomes. Meta-analytic evidence indicates that, compared with extrinsic motivation, intrinsic motivation has

---

[89] Timothy A. Judge, Chad A. Higgins, Carl J. Thoresen, and Murray R. Barrick, "The Big Five Personality Traits, General Mental Ability, and Career Success Across the Life Span," *Personnel Psychology*, Vol. 52, No. 3, 1999, pp. 621–652. As with other "Big Five" personality characteristics, conscientiousness has different dimensions depending on the measure and model. For details on these differences, see Joshua J. Jackson, Tim Bogg, Kate E. Walton, Dustin Wood, Peter D. Harms, Jennifer Lodi-Smith, Grant W. Edmonds, and Brent W. Roberts, "Not All Conscientiousness Scales Change Alike: A Multimethod, Multisample Study of Age Differences in the Facets of Conscientiousness," *Journal of Personality and Social Psychology*, Vol. 96, No. 2, 2009, pp. 446–459.

[90] On job performance, see Murray R. Barrick, and Michael K. Mount, "The Big Five Personality Dimensions and Job Performance: A Meta-Analysis," *Personnel Psychology*, Vol. 4, No. 1, 1991, pp. 1–26. See also Judge et al., 1999. Facets of conscientiousness also predict outcomes outside the work setting, including health behaviors and life outcomes (e.g., divorce). For example, for evidence of how higher conscientiousness predicts lower engagement in risky health behaviors (e.g., tobacco use), see Timothy Bogg and Brent W. Roberts, "Conscientiousness and Health-Related Behaviors: A Meta-Analysis of the Leading Behavioral Contributors to Mortality," *Psychological Bulletin*, Vol. 130, No. 6, 2004, pp. 887–919.

[91] Grant W. Edmonds, Joshua J. Jackson, Jennifer V. Fayard, and Brent W. Roberts, "Is Character Fate, or Is There Hope to Change My Personality Yet?" *Social and Personality Psychology Compass*, Vol. 2, No. 1, 2008, pp. 399–413.

[92] Susan G. Straus, Michael G. Shanley, Carra S. Sims, Bryan W. Hallmark, Anna R. Saavedra, Stoney Trent, and Sean Duggan, *Innovative Leader Development: Evaluation of the U.S. Army Asymmetric Warfare Adaptive Leader Program*, Santa Monica, Calif.: RAND Arroyo Center, RR-504-A, 2014.

[93] Consulting with DAF psychologists is strongly recommended to ensure that any personality assessment selected is valid, results are interpreted accurately, and feedback provided is appropriate.

[94] Terence R. Mitchell and Denise Daniels, "Motivation," in Walter C. Borman, Daniel R. Ilgen, and Richard J. Klimoski, eds., *Handbook of Psychology*, Vol. 12: *Industrial and Organizational Psychology*, Hoboken, N.J.: John Wiley & Sons, 2003, pp. 225–254.

stronger relationships with work-related outcomes, including job attitudes (e.g., job satisfaction) and behavior (e.g., lower engagement in counterproductive work behaviors).[95]

Other key concepts tie into work-motivation theories. We highlight three concepts that have received significant attention in research on work motivation: self-regulation, goal-setting, and self-efficacy.

- **Self-regulation** refers to an individual's ability to control or *regulate* their emotions, thoughts, and behaviors to engage in activities to achieve goals. Self-regulation is a central concept in several work-motivation theories.[96]
- **Goal-setting** theory has as its central tenant that "specific, high (hard) goals lead to a higher level of task performance than do easy goals or vague, abstract goals such as the exhortation to 'do one's best.'"[97] Goal-setting theory has been supported by several studies and meta-analyses. For example, one recent meta-analytic study examined predictors of work effort.[98] It found that goal-setting (as in the act of setting goals, not the full theory) has a small-to-medium-size relationship with work effort (mean corrected correlation = 0.32) while goal commitment has a medium-to-strong relationship with work effort (mean corrected correlation = 0.47).
- **Self-efficacy** refers to a person's belief in their capabilities to be successful at achieving certain goals.[99] Those with high self-efficacy tend to set more challenging performance goals and exert effort to meet those goals, thus enhancing performance.[100] A recent meta-analysis shows that self-efficacy has a moderately strong relationship with work effort (mean corrected correlation of 0.40).[101]

In addition to factors centered on the individual (e.g., self-regulation, personality, ability), features of the job and work environment affect work effort, too. Specifically, the meta-analysis of work effort relationships finds that job and organizational factors have comparably sized correlations with work effort as individual-focused factors like conscientiousness and intrinsic motivation.[102] In particular, the variety of skills that workers employ on the job (skill variety), the importance of the work to the organization or society (task significance), and perceived fit between the person and the job (person-

---

[95] Anja Van den Broeck, Joshua L. Howard, Yves Van Vaerenbergh, Hannes Leroy, and Marylene Gagne, "Beyond Intrinsic and Extrinsic Motivation: A Meta-Analysis on Self-Determination Theory's Multidimensional Conceptualization of Work Motivation," *Organizational Psychology Review*, Vol. 11, No. 3, 2021, pp. 240–273.

[96] Jeffrey B. Vancouver, and David V. Day, "Industrial and Organisation Research on Self-Regulation: From Constructs to Application," *Applied Psychology: An International Review*, Vol. 54, No. 2, 2005, pp. 155–185.

[97] Edwin A. Locke and Gary P. Latham, "New Directions in Goal-Setting Theory," *Current Directions in Psychological Science*, Vol. 15, No. 5, 2006, p. 265. See also Edwin A. Locke and Gary P. Latham, *A Theory of Goal Setting and Task Performance*, Englewood Cliffs, N.J.: Prentice-Hall, 1990; Edwin A. Locke and Gary P. Latham, "Building a Practically Useful Theory of Goal Setting and Task Motivation: A 35-Year Odyssey," *American Psychologist*, Vol. 57, 2002, pp. 705–717.

[98] Van Iddekinge et al., 2022.

[99] Robert Wood and Albert Bandura, "Impact of Conceptions of Ability on Self-Regulatory Mechanisms and Complex Decision Making," *Journal of Personality and Social Psychology*, Vol. 56, No. 3, 1989, pp. 407–415.

[100] Albert Bandura, and Richard H. Walters, *Social Learning Theory*, Vol. 1: Englewood Cliffs, N.J.: Prentice Hall, 1977; Albert Bandura, "On the Functional Properties of Perceived Self-Efficacy Revisited," *Journal of Management*, Vol. 28, 2012, pp. 9–44; Albert Bandura and Edwin A. Locke, "Negative Self-Efficacy and Goal Effects Revisited," *Journal of Applied Psychology*, Vol. 88, 2003, pp. 87–99.

[101] Van Iddekinge et al., 2022.

[102] Van Iddekinge et al., 2022.

job fit) had sizable correlations with work effort (mean corrected correlations of 0.49, 0.41, and 0.46, respectively). Other theories of motivation emphasize individual perceptions of the relationships between effort, performance, and expected outcomes. That is, effort is expected to be lower when students perceive that increased effort will not lead to higher performance. The perceived relationships between effort and performance do not always reflect reality, especially if students lack knowledge or have a low self-concept.[103]

Evidence-Based Practices for Developing Drive

Although changing personality characteristics like conscientiousness may not be particularly successful, there are work-motivation interventions that could improve Drive. Table 6.2 lists interventions that have received empirical support in scholarly literature.

Table 6.2. Evidence-Based Interventions for Developing Drive

Intervention	Description
Goal-setting interventions	While there is no single format for goal-setting interventions (which tend to be integrated with other workplace training), there are two main features: (1) participants set specific but challenging goals aligned with performing one or more tasks, and (2) participants commit to the goal. Such interventions are often accompanied by feedback sessions and assessments of participant self-efficacy.
Self-management training	Generally, this training includes several phases to teach participants how to do the following: self-assess which behaviors to change; set goals aligned with the behaviors the individual wishes to change; monitor one's own progress toward the selected goal(s); evaluate oneself and use reinforcing or punishing techniques; draft a contract with oneself on the goals, expectations, and outcomes should goals not be met; and maintain progress by overcoming common obstacles.[a]
Specific interventions focused on increasing enjoyment of goal-directed activities	Evidence suggests that enjoying activities associated with goals can increase intrinsic motivation toward goal attainment. Examples of interventions to increase goal-directed activities include the following: • Offer choice in how to achieve the goal (e.g., if goal is to increase physical fitness, offer choices for the types of exercises people can do to achieve the goal). • Provide benefits during or immediately after performing the activities (e.g., "bundle" something you want to do [watch a TV show] with something you ought to do [exercise] to achieve a fitness goal). • Place focus on the positive experience of performing the activities, not on the output of the activities (e.g., focusing on the experience of running rather than the goal of achieving a certain run time or distance). This may involve techniques such as mindfulness training, "nostalgic reflections" (e.g., reminiscing about performing those activities in the past while performing those activities in the present), and promoting a sense of "togetherness" when performing team activities.[b]

[103] Ruth Kanfer and Phillip L. Ackerman, "Aging, Adult Development, and Work Motivation," *Academy of Management Review*, Vol 29, 2004, pp. 440–458.

Intervention	Description
Providing incentives	External incentives, such as awards, can increase motivation; however, they tend to work better when the performance outcome is measured in terms of quantity (e.g., number of events completed) rather than quality.[c]

NOTES:
[a] See Colette A. Frayne and J. Michael Geringer, "Self-Management Training for Improving Job Performance: A Field Experiment Involving Salespeople," *Journal of Applied Psychology*, Vol. 85, No. 3, 2000, Table 1, p. 362.
[b] Ayelet Fishbach and Kaitlin Woolley, "The Structure of Intrinsic Motivation," *Annual Review of Organizational Psychology and Organizational Behavior*, Vol. 9, 2022, pp. 349–352.
[c] Christopher P. Cerasoli, Jessica M. Nicklin, and Michael T. Ford, "Intrinsic Motivation and Extrinsic Incentives Jointly Predict Performance: A 40-Year Meta-Analysis," *Psychological Bulletin*, Vol. 140, No. 4, 2014, pp. 980–1008; Ji Hyun Kim, Barry Gerhart, and Meiyu Fang, "Do Financial Incentives Help or Harm Performance in Interesting Tasks?" *Journal of Applied Psychology*, Vol. 107, No. 1, 2022, pp. 153–167.

Conclusions

The evidence-based practices in Table 6.2 range from those that are part of formal training interventions involving goal-setting or self-management techniques to those that may be included as part of a regular work environment (not formal training). For the more formal training interventions, the ultimate success will depend on how well the training transfers to the workplace. As noted in Appendix A, transfer of training is affected by different factors, such as a climate that supports training transfer. In general, having supportive leaders and peers, available resources and opportunities to use what was learned in training, and constructive feedback in the workplace are important for training interventions to move from the training environment to the work environment.[104]

Integrity

Attribute Definition and Background

Although scholarly literature has different definitions of Integrity, for ASPECWAR, the definition falls more in line with upholding "a socially accepted code of morality"[105] (see Table 6.3). To succeed in AFSPECWAR, one must uphold the values that AFSPECWAR deems important and act with those values in mind. Definitions like this contain some form of honesty, fairness, and behaving in an ethical way with consistency.

Table 6.3. Integrity Facets and Definitions

Integrity	Behaves in an Honest, Fair, and Ethical Manner
Obligation	Fulfilling duties, following rules, and avoiding misconduct/abuse (e.g., as defined in the UCMJ)
Courage	Overcoming adversity or personal risk (physical or social) to sacrifice for the greater good and go above the call of duty (i.e., virtuous, honorable)

[104] Brian D. Blume, J. Kevin Ford, Timothy T. Baldwin, and Jason L. Huang, "Transfer of Training: A Meta-Analytic Review," *Journal of Management*, Vol. 36, No. 4, 2010, pp. 1065–1105.

[105] Gabrielle Taylor and Raimond Gaita, "Integrity," *Aristotelian Society Supplementary Volume*, Vol. 55, No. 1, 1981, p. 143.

Concepts Related to Integrity

Although there are several concepts related to Integrity as an attribute, we highlight those that are relevant in the workplace. We describe them in terms of concepts that overlap with the definition of integrity, those that describe behavioral outcomes of not having (or having low levels of) integrity, and individual differences associated with integrity.

Overlapping Concepts

The concept of *professionalism* comes up in in a variety of military organizations' lists of attributes that are important for Special Operations Forces (SOF) and paramilitary roles, yet definitions vary.[106] In addition to including integrity as part of its definition, professionalism also includes concepts such as teamwork, politeness, organization, availability, pursuit of excellence, and collaboration.[107] However, the specific values and behaviors can vary by profession, leading to different interpretations of its meaning.[108]

Another concept that overlaps with Integrity is *moral courage*, or a "willingness to speak out and do that which is right in the face of forces that would lead person to act in some other way."[109] Moral courage aligns with Integrity in the sense of behaving ethically in a consistent manner free from outside influence.[110] This perspective aligns with values of respect, empathy, moral reasoning, honesty, and abiding by rules.

Behavioral Outcomes Associated with a Lack of Integrity

The concepts of unethical behavior and counterproductive work behaviors (CWBs) offer behavioral categories associated with low levels of Integrity. Unethical behavior involves "violat[ing] a prescribed norm"; examples are condoning wrongdoing, deception/dishonesty, unfair treatment of others, rules violation, and exploitation/cheating/fraud.[111] CWBs are violations of the interests and

---

[106] Krueger et. al., 2022.

[107] Tim J. Wilkinson, Winnie B. Wade, and L. Doug Knock, "A Blueprint to Assess Professionalism: Results of a Systematic Review," *Academic Medicine*, Vol. 84, No. 5, 2009, pp. 551–558; Cara S. Lesser, Catherine R. Lucey, Barry Egener, Clarence H. Braddock, Stuart L. Linas, and Wendy Levinson. "A Behavioral and Systems View of Professionalism," *Journal of the American Medical Association*, Vol. 304, No. 24, 2010, pp. 2732–2737.

[108] Alexandra Lecours, Nancy Baril, and Marie-Josée Drolet, "What Is Professionalism in Occupational Therapy? A Concept Analysis: Qu'est-ce que le professionnalisme en ergothérapie? Analyse de ce concept," *Canadian Journal of Occupational Therapy*, Vol. 88, No. 2, 2021, pp. 117–130.

[109] Vicki D. Lachman, "Moral Courage: A Virtue in Need of Development?" *MESURG Nursing*, Vol. 16, No. 2, 2007, p. 131.

[110] Bernd Marcus and Heinz Schuler, "Antecedents of Counterproductive Behavior at Work: A General Perspective," *Journal of Applied Psychology*, Vol. 89, No. 4, 2004, pp. 647–660; Christopher M. Berry, Deniz S. Ones, and Paul R. Sackett, "Interpersonal Deviance, Organizational Deviance, and Their Common Correlates: A Review and Meta-Analysis," *Journal of Applied Psychology*, Vol. 92, No. 2, 2007, pp. 410–424.

[111] Teresa L. Russell, Taylor E. Sparks, John P. Campbell, Kristina Handy, Peter Ramsberger, and James A. Grand, "Situating Ethical Behavior in the Nomological Network of Job Performance," *Journal of Business and Psychology*, Vol. 32, No. 3, 2017, p. 254; Brenton M. Wiernik and Deniz S. Ones, "Ethical Employee Behaviors in the Consensus Taxonomy of Counterproductive Work Behaviors," *International Journal of Selection and Assessment*, Vol. 26, No. 1, 2018, pp. 36–48.

goals of an organization.[112] Examples of CWBs include theft, fraud, absenteeism, physical and verbal aggression, or substance use. Those with Integrity are less prone to commit CWBs.[113]

Individual Differences Associated with Integrity

Certain personality characteristics have been shown to be associated with Integrity or behavioral outcomes like CWBs that are indicative of low levels of Integrity. Conscientiousness (being reliable, controlled, orderly, achievement oriented) and agreeableness (being cooperative, trusting, and good-natured) are associated with different types of CWB.[114] Self-control, a facet of conscientiousness, may be particularly relevant.[115] Self-control involves avoiding acting on one's impulses and considering longer-term consequences of one's actions. Having self-control can help individuals avoid problematic behaviors like CWBs because they consider the negative consequences (e.g., punishment) of engaging in those behaviors, even if there may be immediate gratification from engaging in the behavior (e.g., enjoyment from having access to a stolen item).

In addition to Big Five personality constructs related to Integrity, the Honesty–Humility trait from the HEXACO personality model has also been found to predict CWBs.[116] Honesty–Humility is defined as "the tendency to be fair and genuine in dealing with others, in the sense of cooperating with others even when one might exploit them without suffering retaliation."[117] Honesty–Humility was also found to be highly correlated with measures of integrity.[118]

Besides personality, a person's cognitive ability may also relate to Integrity by reducing the risk of engaging in CWBs. The mechanism through which cognitive ability is connected to unethical behavior may be through moral reasoning "as it entails recognizing the implicit rules underlying civilized interaction."[119] Although certain personality characteristics and cognitive ability may relate to integrity, they tend not to be particularly amenable to change through training intervention.[120] Other types of interventions may be more effective and are discussed in the next section.

---

[112] Marcus and Schuler, 2004.

[113] Berry, Ones, and Sackett, 2007.

[114] Paul E. Spector, "The Relationship of Personality to Counterproductive Work Behavior (CWB): An Integration of Perspectives," *Human Resource Management Review*, Vol. 21, No. 4, 2011, pp. 342–352.

[115] Brent W. Roberts, Oleksandr S. Chernyshenko, Stephen Stark, and Lewis R. Goldberg, "The Structure of Conscientiousness: An Empirical Investigation Based on Seven Major Personality Questionnaires," *Personnel Psychology*, Vol. 58, No. 1, 2005, pp. 103–139.

[116] The HEXACO model describes six dimensions of personality rather than five. This model stemmed from lexical studies of personality structure, examining the words used to describe personality in many different languages. Reviews of these studies found six dimensions of personality, Honesty–Humility, Emotionality, Extraversion, Agreeableness, Conscientiousness, and Openness to Experience. See Kibeom Lee, Michael C. Ashton, and Reinout E. de Vries, "Predicting Workplace Delinquency and Integrity with the HEXACO and Five-Factor Models of Personality Structure," *Human Performance*, Vol. 18, No. 2, 2005, 179–197.

[117] Michael C. L. Ashton, and Kibeom Lee, "Empirical, Theoretical, and Practical Advantages of the HEXACO Model of Personality Structure," *Personality and Social Psychology Review*, Vol. 11, No. 2, 2007, p. 156.

[118] Lee, Ashton, and de Vries, 2005, 179–197.

[119] Stephan Dilchert, Deniz S. Ones, Robert D. Davis, and Cary D. Rostow, "Cognitive Ability Predicts Objectively Measured Counterproductive Work Behaviors," *Journal of Applied Psychology*, Vol. 92, No. 3, 2007, p. 618.

[120] David Moreau, "How Malleable Are Cognitive Abilities? A Critical Perspective on Popular Brief Interventions," *American Psychologist*, Vol. 77, No. 3, 2022, pp. 409–423.

Evidence-Based Practices for Integrity

Effective practices for Integrity in the workplace focus on promoting ethical behavior. That is, the focus is not on changing someone's personality, values, or other relatively stable characteristics associated with Integrity. Instead, the focus is on teaching ethical behaviors and creating conditions that promote the behavior in the work environment.

Ethics training is a common tool used by organizations and in academic settings. Evidence for the effectiveness of ethics training is mixed overall. However, certain characteristics of ethics training may be effective. Specifically, ethics training may be more effective when it involves use of the following:

- cases or scenarios that present ethical problems[121]
- experiential learning methods that emotionally engage participants with the ethical problems they are asked to address[122]
- individually based training[123]
- content specific to the field to which the ethical scenarios apply (e.g., relevant to AFSPECWAR environment) as opposed to using a mix of field-specific and field-agnostic content.[124]

Perhaps more effective than formal ethics training in the workplace is creating conditions in the work environment that promote a culture of Integrity.[125] Ethical leadership is particularly important. Ethical leadership involves "the demonstration of normatively appropriate conduct through personal actions and interpersonal relationships, and the promotion of such conduct to followers through two-

---

[121] See, for example, Ethan P. Waples, Alison L. Antes, Stephen T. Murphy, Shane Connelly, and Michael D. Mumford, "A Meta-Analytic Investigation of Business Ethics Instruction," *Journal of Business Ethics*, Vol. 87, No. 1, 2009, pp. 133–151; Jason Borenstein, Matthew J. Drake, Robert Kirkman, and Julie L. Swann, "The Engineering and Science Issues Test (ESIT): A Discipline-Specific Approach to Assessing Moral Judgment," *Science and Engineering Ethics*, Vol. 16, No. 2, 2010; Carole L. Jerkiewicz, "The Phantom Code of Ethics and Public Sector Reform," *Journal of Public Affairs*, Vol. 6, No. 3, May 2002; Christopher H. Warner, George N. Appenzeller, Angela Mobbs, Jessica R. Parker, and Carolynn M. Warner, "Effectiveness of Battlefield-Ethics Training During Combat Deployment: A Programme Assessment," *Lancet*, Vol. 378, 2011; and Johannes Katsarov, Roberto Andorno, André Krom, and Mariëtte van den Hoven, "Effective Strategies for Research Integrity Training—a Meta-analysis," *Educational Psychology Review*, 2021, pp. 1–21.

[122] Katsarov et al., 2021.

[123] Logan L. Watts, Kelsey E. Medeiros, Tyler J. Mulhearn, Logan M. Steele, Shane Connelly, and Michael D. Mumford, "Are Ethics Training Programs Improving? A Meta-Analytic Review of Past and Present Ethics Instruction in the Sciences," *Ethics & Behavior*, Vol. 27, No. 5, 2017, 351–384; William B. Brinkman, Sheela R. Geraghty, Bruce P. Lanphear, Jane C. Khoury, Javier A. Gonzalez del Rey, Thomas G. DeWitt, Maria T. Britto, "Effect of Multisource Feedback on Resident Communication Skills and Professionalism: A Randomized Controlled Trial," *Archives of Pediatrics and Adolescent Medicine*, Vol. 161, January 2007.

[124] Torrence et al. indicates that domain- or field-specific training "has proven to be an effective approach as it familiarizes trainees with the norms and rules guiding responsible behavior within their discipline" (Brett S. Torrence, Logan L. Watts, Tyler J. Mulhearn, Megan R. Turner, E. Michelle Todd, Michael D. Mumford, and Shane Connelly, "Curricular Approaches in Research Ethics Education: Reflecting on More and Less Effective Practices in Instructional Content," *Accountability in Research*, Vol. 24, No. 5, 2017, p. 280).

[125] Kayes et al. describes four characteristics of a culture of integrity: (1) the language of ethical decisionmaking is used; (2) structural supports and procedures that facilitate ethical decisionmaking have been developed; (3) a culture of openness, responsibility, and commitment to multiple organizational goals has been created and sustained; and (4) subordinate development is valued (D. Christopher Kayes, David Stirling, and Tjai M. Nielson, "Building Organizational Integrity," *Business Horizons*, Vol. 50, 2007, p. 63).

way communication, reinforcement, and decision-making."[126] A meta-analysis by Bedi et al. shows that ethical leadership is positively associated with job satisfaction, organizational commitment, organizational citizenship behaviors, job performance, and job engagement in followers.[127] They also found ethical leadership to be negatively associated with CWBs. Ethical leadership can be demonstrated by role modeling, communicating the importance of behaving ethically, and offering rewards for certain ethical behaviors (and punishment for ethical transgressions).[128]

Studies have shown that feedback and accountability can help improve ethical behaviors in leadership; other studies demonstrated that feedback from followers led to improved behavioral performance.[129] Multisource feedback has also been shown to be positively related to professionalism in the workplace.[130] These feedback systems can help leaders reflect on how their behaviors affect subordinates and allow organizations to reinforce ethical behavior through promotion or compensation.

Conclusions

Based on our review, developing integrity should focus on practices to promote ethical behavior. Key strategies include the following:

- Promote ethical behavior through role modeling, effective accountability mechanisms, and regular communication on the importance of desired ethical behaviors.
- Incorporate multisource feedback to hold leaders accountable for their behaviors.
- Use ethical-based scenarios that are specific to the environment in which operators are likely to face ethical dilemmas.

Stress Tolerance

Knowing how to manage stress is a fundamental skill for all military personnel, regardless of service branch, occupation, or deployment status. Stress has an immediate and direct effect on task performance, and it can have both short- and long-term effects on emotional and physical health. For example, a systematic review of Stress Tolerance training focused on long-term outcomes such as

---

[126] Michael E. Brown, Linda K. Treviño, and David A. Harrison, "Ethical Leadership: A Social Learning Perspective for Construct Development and Testing," *Organizational Behavior and Human Decision Processes*, Vol. 97, No. 2, 2005, pp. 117–134.

[127] Akanksha Bedi, Can M. Alpaslan, and Sandy Green, "A Meta-Analytic Review of Ethical Leadership Outcomes and Moderators," *Journal of Business Ethics*, Vol. 139, No. 3, 2016, pp. 517–536.

[128] Linda Klebe Treviño, Laura Pincus Hartman, and Michael Brown, "Moral Person and Moral Manager: How Executives Develop a Reputation for Ethical Leadership," *California Management Review*, Vol. 42, No. 4, 2000, p. 120.

[129] Leanne E. Atwater, Joan F. Brett, and Atira Cherise Charles, "Multisource Feedback: Lessons Learned and Implications for Practice," *Human Resource Management*, Vol. 46, No. 2, 2007; Leanne Atwater, Paul Roush, and Allison Fischthal, "The Influence of Upward Feedback on Self- and Follower Ratings of Leadership," *Personnel Psychology*, Vol. 48, No. 1, 1995.

[130] Brinkman et al., 2007; Jinwei Hu, Robert Lee, Sarah Mullin, Steven Schwaitzberg, Larry Harmon, Paul Gregory, and Peter L. Elkin, "How Physicians Change: Multisource Feedback Driven Intervention Improves Physician Leadership and Teamwork," *Surgery*, Vol. 168, 2020.

posttraumatic stress disorder (PTSD), depression, anxiety, misuse of alcohol and other substances, and marriage/family outcomes and medium-term outcomes such as absenteeism and return to duty.[131] Although these outcomes are relevant for any military community,[132] AFSPECWAR training emphasizes the importance of employing technical skills in dynamic and difficult conditions. Thus, Stress Tolerance is a key attribute to success in the training pipeline and in operational environments.

Attribute Definition and Background

In AFSPECWAR, Stress Tolerance allows operators to have the composure to communicate effectively during extremely volatile situations as well as maintain cognitive function (attention, memory) under time pressure, substantial uncertainty, and constant distraction. Table 6.4 presents the definition and facets of Stress Tolerance.

Table 6.4. Stress Tolerance Facets and Definitions

Stress Tolerance	**Continuing to Perform Under Difficult, Unpredictable, and Ambiguous Conditions**
Emotion Regulation	Managing verbal (e.g., shouting, complaining) and nonverbal expressions (e.g., eye-rolling, head down)
Performance (Under Stress)	Continuing to perform and meet the demands of the task (e.g., not freezing)

Because this focused definition of Stress Tolerance emphasizes an individual's *immediate* response to a clearly defined stressor, we do not include stress management efforts to achieve other important outcomes, such as prevention of PTSD, improved mental health, and increased physical health.

To identify evidence-based practices for developing Stress Tolerance, their common elements, and future directions for development, we conducted a targeted search for relevant literature. We describe our search strategy in Appendix C. Overall, the research in this area is weak. Studies were often conducted by the program developers who have an inherent conflict of interest in reporting results. Often outcomes were self-reported by trainees or collected using unvalidated instruments created by the program developers. Some studies had no control or comparison group, so improvements in tolerating stress may have occurred for reasons other than the specific training program, such as changes in the environment due to deployment conditions. Other than a few reports on mental health issues such as PTSD or substance use disorder, studies with military populations did not measure the long-term (more than six months after training) effects of the training on emotional regulation or stress tolerance.

[131] Margaret A. Maglione, Christine Chen, Armenda Bialas, Aneesa Motala, Joan Chang, Goke Akinniranye, and Susanne Hempel, *Stress Control for Military, Law Enforcement, and First Responders: A Systematic Review*, Santa Monica, Calif.: RAND, RR-A119-3, 2021.

[132] Sarah A. McInerney, Edward Waldrep, and Charles C. Benight, "Resilience Enhancing Programs in the U.S. Military: An Exploration of Theory and Applied Practice," *Military Psychology*, 2022, pp. 1–12.

Approaches to Develop Stress Tolerance

There are two broad approaches designed to increase Stress Tolerance:[133]

- Resilience training programs teach stress management, coping techniques and psychological skills through education and practice.[134] Although resilience training can target emotion regulation and support performance under stress, interventions are often more broadly designed to prevent negative longer-term outcomes such as PTSD, anxiety disorder, and major depressive disorder.
- Stress Inoculation Training (SIT) aims to build Stress Tolerance through direct exposure to acute stressors after psychological skills are taught. SIT is based on the theory that developing a healthy stress reaction requires an initial exposure to stress, which has an inoculating effect.[135] Situations that are stressful enough to challenge but not overwhelm provide an opportunity to learn skills or identify strategies that can help the individual to overcome future demands.

Both SIT and resilience training teach psychological skills and strategies to reduce anxiety and enhance one's ability to respond effectively during stressful circumstances; however, there is no consensus on which psychological skills are most effective. Consequently, a range of different skills may be taught, and then students determine what works best for them in a particular situation.

Evidence-Based Practices for Developing Stress Tolerance

In the following sections, we describe the evidence for different types of psychological skills training, resilience training, and conclude with a more in-depth discussion of SIT, which has demonstrated the most promising results.

---

[133] Researchers have employed diverse outcome measures to study the effect of programs designed to increase Stress Tolerance. Emotional stability has been measured on scales assessing optimism (versus pessimism), anxiety, aggression, dependence, and empathy. Physiological measures like heart rate, blood pressure, and salivatory cortisol have been used to estimate stress levels during simulated combat situations. Performance during stressful military tasks has been assessed using standardized measures of attention and working memory, in addition to task evaluation by instructors. Finally, scales measuring attributes like self-efficacy, perseverance, life purpose, and negative affect have been used to assess resilience. Three scales have the promising psychometric properties: the Connor-Davidson Resilience Scale (CD-RISC), the Resilience Scale for Adults, and the Brief Resilience Scale (BRS). See Kathryn M. Connor and Jonathan R. T. Davidson, "Development of a New Resilience Scale: The Connor-Davidson Resilience Scale (CD-RISC)," *Depression and Anxiety*, Vol. 18, No. 2, 2003. For adult resilience, see Oddgeir Friborg, Odin Hjemdal, Jan H. Rosenvinge, and Monica Martinussen, "A New Rating Scale for Adult Resilience: What Are the Central Protective Resources Behind Healthy Adjustment?" *International Journal of Methods in Psychiatric Research*, Vol. 12, No. 2, 2003; for BRS, see Bruce W. Smith, Jeanne Dalen, Kathryn Wiggins, Erin Tooley, Paulette Christopher, and Jennifer Bernard, "The Brief Resilience Scale: Assessing the Ability to Bounce Back," *International Journal of Behavioral Medicine*, Vol. 15, No. 3, 2008.

[134] Lisa S. Meredith, Cathy D. Sherbourne, Sarah J. Gaillot, Lydia Hansell, Hans V. Ritschard, Andrew M. Parker, and Glenda Wrenn, *Promoting Psychological Resilience in the U.S. Military*, Santa Monica, Calif.: RAND, MG-996-OSD, 2011.

[135] Meredith et al., 2011; Daniel Birrer and Gareth Morgan, "Psychological Skills Training as a Way to Enhance an Athlete's Performance in High-Intensity Sports," *Scandinavian Journal of Medicine & Science in Sports*, Vol. 20, Supp. 2, 2010, pp. 78–87.

Psychological Skills Training

An important component of interventions to promote Stress Tolerance is the development of specific skills to manage stress. Although much more research is needed on the benefits of specific psychological skills,[136] especially within the context of performing under stress, strategies with some supporting evidence include self-talk, mental practice, imagery, and adopting a growth mindset:

- Self-talk strategy training produces performance gains in athletes.[137] Self-talk was found to be more useful when first learning tasks, when instructional versus motivational cues were used, and for fine versus gross motor tasks.
- Mental practice and imagery training have been found to be effective strategies for improving performance.[138] Mental practice is the cognitive rehearsal of a motor movement without overt physical movements; it does not necessarily involve imagery and can refer to other types of mental preparation.[139] In contrast, imagery training is the specific mental creation and re-creation of any kind of experience; it is not limited to motor movement.[140]
- Mindset, a core assumption that influences how an individual reacts and responds to different situations, has been found to affect performance.[141] Mindsets surrounding stress typically are categorized in two ways: a belief that stress enhances health, and a belief that stress has debilitating consequences.[142] A recent study found that Navy SEALs who believed stress is enhancing demonstrated greater persistence through training, faster obstacle course times, and fewer negative evaluations from peers and instructors.[143] Many current Stress Tolerance training programs emphasize stress-management and stress-reduction techniques.[144] Few of them examine how mindset training can influence performance. However, there is some preliminary evidence that individuals can be trained to develop stress is enhancing mindsets.[145]

---

[136] Birrer and Morgan, 2010.

[137] Antonis Hatzigeorgiadis, Nikos Zourbanos, Evangelos Galanis, and Yiannis Theodorakis, "Self-Talk and Sports Performance: A Meta-Analysis," *Perspectives on Psychological Science*, Vol. 6, No. 4, 2011, pp. 348–356.

[138] Bianca A. Simonsmeier, Melina Andronie, Susanne Buecker, and Cornelia Frank, "The Effects of Imagery Interventions in Sports: A Meta-Analysis," *International Review of Sport and Exercise Psychology*, Vol. 14, No. 1, 2021, pp. 186–207; Adam J. Toth, Eoghan McNeill, Kevin Hayes, Aidan P. Moran, and Mark Campbell, "Does Mental Practice Still Enhance Performance? A 24 Year Follow-Up and Meta-Analytic Replication and Extension," *Psychology of Sport and Exercise*, Vol. 48, 2020, pp. 1–13.

[139] Toth et al., 2020.

[140] Simonsmeier et al., 2021.

[141] Carol S. Dweck, *Mindset: The New Psychology of Success*, New York: Random House, 2006.

[142] Alia J. Crum, Modupe Akinola, Ashley Martin, and Sean Fath, "The Role of Stress Mindset in Shaping Cognitive, Emotional, and Physiological Responses to Challenging and Threatening Stress," *Anxiety, Stress, & Coping*, Vol. 30, No. 4, 2017, pp. 379–395.

[143] Eric N. Smith, Michael D. Young, and Alia J. Crum, "Stress, Mindsets, and Success in Navy SEALs Special Warfare Training," *Frontiers in Psychology*, Vol. 10, 2020.

[144] Maglione et al., 2021; Federica Pallavicini, Luca Argenton, Nicola Toniazzi, Luciana Aceti, and Fabrizia Mantovani, "Virtual Reality Applications for Stress Management Training in the Military," *Aerospace Medicine and Human Performance*, Vol. 87, No. 12, 2016, pp. 1021–1030; George T. Patterson, Irene W. Chung, and Philip W. Swan, "Stress Management Interventions for Police Officers and Recruits: A Meta-Analysis," *Journal of Experimental Criminology*, Vol. 10, No. 4, 2014, pp. 487–513.

[145] Maglione et al., 2021; Alia Crum, "Evaluating a Mindset Training Program to Unleash the Enhancing Nature of Stress," in *Academy of Management Proceedings*, Vol. 2011, No. 1, Briarcliff Manor, N.Y.: Academy of Management, 2011, pp. 1–6.

- Pressure training involves performers practicing physical or technical skills under simulated psychological pressure. Although interventions are not described in depth, Low et al.'s systematic review of pressure training for performance among law enforcement officers and athletes found that training with simulated consequences had a positive effect.[146] Specifically, pressure was increased by adding monetary rewards, punishments, or perceived evaluation by coaches.[147]

Resilience Training

Resilience training interventions involve the instruction and practice of stress coping mechanisms while in a nonstressful setting. Reviews of programs for the Department of Defense and the USAF to promote resilience suggested the use of techniques such as adopting positive coping strategies, maintaining optimism, restructuring problems in a more positive or realistic way, and enhancing self-regulation.[148] However, other evaluations among first responders have highlighted the need to advance the development of resilience training programs for personnel exposed to acute, prolonged stress within their line of work. Patterson, Chung, and Swan conducted a systematic review and meta-analysis of 12 stress-management interventions for law enforcement.[149] All joint analysis of the interventions indicated no significant physiological, psychological, or behavioral outcomes. The authors emphasized the need for the development and rigorous evaluation of stress-management interventions that addressed specific, field-relevant stressors. Similarly, in a rapid overview of systematic reviews of programs designed to prevent and manage occupational stress among first responders, Anthony et al. cautioned that gaps in the design and evaluation of interventions should be addressed before widespread implementation.[150]

One interesting place to begin addressing gaps is imagery rehearsal training. Both a rapid overview and the systematic review of stress interventions in military personnel highlighted one small, randomized controlled trial of imagery training with law enforcement.[151] The training combined psychoeducation with skill development in imagery training and relaxation techniques. Results indicated a positive effect on emotional stability and task performance in officers. Further study of imagery rehearsal in military contexts could provide important data.

---

[146] William R. Low, G. R. H. Sandercock, Paul Freeman, Marie E. Winter, Joanne Butt, and Ian Maynard, "Pressure Training for Performance Domains: A Meta-Analysis," *Sport, Exercise, and Performance Psychology*, Vol. 10, No. 1, 2021, pp. 149–163.

[147] On monetary rewards, see Raôul R. D. Oudejans, "Reality-Based Practice Under Pressure Improves Handgun Shooting Performance of Police Officers," *Ergonomics*, Vol. 51, 2008, pp. 261–273. On the use of punishments, see James J. Bell, Lew Hardy, and Stuart Beattie, "Enhancing Mental Toughness and Performance Under Pressure in Elite Young Cricketers: A 2-Year Longitudinal Intervention," *Sport, Exercise, and Performance Psychology*, Vol. 2, 2013, pp. 281–297. On the evaluations by coaches, see Bradley Beseler, Christopher Mesagno, Warren Young, and Jack Harvey, "Igniting the Pressure Acclimatization Training Debate: Contradictory Pilot-Study Evidence from Australian Football," *Journal of Sport Behavior*, Vol. 39, 2016, pp. 22–38.

[148] Meredith et al., 2011; Sean Robson, *Psychological Fitness and Resilience: A Review of Relevant Constructs, Measures, and Links to Well-Being*, Santa Monica, Calif.: RAND, RR-102-AF, 2014.

[149] Patterson, Chung, and Swan, 2014.

[150] Jesmin Antony, Raman Brar, Paul A. Khan, Marco Ghassemi, Vera Nincic, Jane P. Sharpe, Sharon E. Straus, and Andrea C. Tricco, "Interventions for the Prevention and Management of Occupational Stress Injury in First Responders: A Rapid Overview of Reviews," *Systematic Reviews*, Vol. 9, No. 1, 2020, pp. 1–20.

[151] For the rapid overview, see Antony et al. 2020. For the systematic overview, see Maglione et al., 2021.

Still, despite the promise of imagery training, no conclusive evidence supports the recommendation of resilience training programs to improve tolerance for stress—in terms of emotional regulation and increased performance—among first responders and military personnel.

Stress Inoculation Training

SIT can be described as a process that starts with a conceptual understanding of the types of stress faced and how individuals have responded to these conditions in the past. After reflecting on what has been effective or ineffective in the past, trainers begin teaching different strategies to effectively manage stress. After learning about these strategies, individuals are given opportunities to practice while being exposed to relevant stressors. Finally, individuals are encouraged to evaluate how effective they were in applying the strategies.

There are several important components to SIT in the context of career fields that have to perform under stressful conditions. Trainees must be able to perform the task and receive feedback on performance under conditions of low stress. Once competence has been demonstrated, trainees can then progress to performing the task under more realistic, operational conditions. Increasing the realism may include the introduction of difficult environmental conditions (e.g., darkness, temperature extremes), psychological conditions (e.g., isolation, interpersonal conflict), or modified general work conditions (e.g., increased urgency, increased number of tasks). In general, "the more realistic the practice setting (the stress exposure), the more effective the method should be."[152] In a report for AFSPECWAR, Robson and Manacapilli describe SIT in more detail and its potential benefits.[153] Although SIT programs have reported promising results, additional studies are needed to determine if findings can be reproduced and generalized to AFSPECWAR.

SIT, specifically training using modern technology, has the most evidence of efficacy in improving Stress Tolerance. These programs use virtual reality or computer video to simulate high-stress military scenarios where trainees can practice learned skills such as tactical breathing. Biofeedback is often provided to trainees during and after simulations. This type of training resulted in lower anxiety, measured by objective biological measures, during live "in-person" simulated scenarios. Other stress inoculation programs report promising results, but whether they have long-term effects on emotional regulation and task performance is unknown.[154]

---

[152] Michael R. Baumann, Carol L. Gohm, and Bryan L. Bonner, "Phased Training for High-Reliability Occupations: Live-Fire Exercises for Civilian Firefighters," *Human Factors*, Vol. 53, No. 5, 2011, p. 558.

[153] Sean Robson and Thomas Manacapilli, *Enhancing Performance Under Stress: Stress Inoculation Training for Battlefield Airmen*, Santa Monica, Calif.: RAND, RR-750-AF, 2014.

[154] For descriptions of these various programs, see Laurel Hourani, Stephen Tueller, Paul Kizakevich, Gregory Lewis, Laura Strange, Belinda Weimer, Stephanie Bryant, Ellen Bishop, Robert Hubal, and James Spira, "Toward Preventing Post-Traumatic Stress Disorder: Development and Testing of a Pilot Predeployment Stress Inoculation Training Program," *Military Medicine*, Vol. 181, No. 9, 2016, pp. 1151–1160; Laurel Hourani, Stephen Tueller, Paul Kizakevich, Laura Strange, Gregory Lewis, Belinda Weimer, Jessica Morgan, Darryl Cooney, and Jessica Nelson, "Effect of Stress Inoculation Training with Relaxation Breathing on Perceived Stress and Posttraumatic Stress Disorder in the Military: A Longitudinal Study," *International Journal of Stress Management*, Vol. 25, No. S1, 2018, pp. 124–136; Gregory F. Lewis, Laurel Hourani, Stephen Tueller, Paul Kizakevich, Stephanie Bryant, Belinda Weimer, and Laura Strange, "Relaxation Training Assisted by Heart Rate Variability Biofeedback: Implication for a Military Predeployment Stress Inoculation Protocol," *Psychophysiology*, Vol. 52, No. 9, 2015, pp. 1167–1174; Stéphane Bouchard, François Bernier, Éric Boivin, Brian Morin, and Geneviève Robillard, "Using Biofeedback While Immersed

Guidelines for Stress Tolerance Training

- In general, successful training programs should first teach concepts and skills relevant to Stress Tolerance, then allow trainees to practice skills in increasingly stressful and realistic conditions. That is, students should learn both the technical content and skills of their jobs and psychological skills before exposure to stressful conditions.
- When teaching psychological skills, ensure that the training material has been tailored to the audience and allows time for discussion and practice.[155]
- Gradually increase realism and stress of practice scenarios, and include multiple training sessions.
- Following exposure to stressful conditions, give trainees time to reflect on how effective they were in applying strategies. Provide feedback on application of skills, either through biofeedback or personalized instruction.
- Continue to evaluate Stress Tolerance development programs. When possible, future research should be conducted by third-party researchers; include a control group, include measurements of task performance and emotional regulation, and measure outcomes at multiple timepoints throughout the training.

Conclusions

Stress Tolerance, the ability to perform under difficult, unpredictable, and ambiguous conditions, can be developed through evidence-based training programs. SIT first educates participants to recognize and manage stress. It then presents opportunities for participants to practice stress management in conditions that approximate the operational environment so that they can identify strategies that are personally effective in optimizing performance under stress. SIT using virtual reality technology has the most evidence of efficacy in improving Stress Tolerance. However, there is no consensus on which techniques and psychological skills are most effective to support SIT. There is some evidence to support self-talk strategy, imagery training, mental practice, attentional control, and diaphragmatic breathing. Rigorous pilot testing should be conducted before any specific program is implemented.

---

in a Stressful Videogame Increases the Effectiveness of Stress Management Skills in Soldiers," *PloS One*, Vol. 7, No. 4, 2012, e36169.

[155] A review of the Air Force Airman Resilience Training program identified lack of engagement and limited emphasis on concrete skill development as factors limiting overall effectiveness; see Gabriella C. Gonzalez, Reema Singh, Terry L. Schell, and Robin M. Weinick, *An Evaluation of the Implementation and Perceived Utility of the Airman Resilience Training Program*, Santa Monica, Calif.: RAND, RR-655-OSD, 2014.

Chapter 7

Developing Interpersonal Attributes

Interpersonal attributes are skills and abilities that are used to relate to others. There are two AFSPECWAR interpersonal attributes: Teamwork and Communication. In this chapter, we review concepts related to these attributes and review best practices for developing them.

Teamwork

For teams to achieve effective outcomes, members must have *taskwork competencies* and *teamwork competencies*. Taskwork competencies are the KSAOs that individuals need to complete the task (e.g., perform water-crossing techniques or coordinate technical infiltration into denied territory). Teamwork competencies are KSAOs that allow team members to interact effectively.[156] In this section, we focus on teamwork competencies as they require team-based training. We summarize research findings on competencies important to SOF teams that are amenable to change through training, and we review research on the effectiveness of methods for training teamwork competencies.

Attribute Definition and Background

There are many models of team competencies positing an abundance of teamwork attributes. Here, we focus on a subset of competencies, in light of the definitions and facets presented in Table 7.1, as well as the findings about factors associated with performance deficiencies in training—exhibited by students who put their own interests before the team or create conflict within the team, as we describe in Chapter 3. In addition, we review training for interpersonal skills or related constructs such as empathy and emotional intelligence.[157] The review of team training methods also provides evidence of specific approaches, such as team building and debriefing, that influence the development of other teamwork competencies, such as trust and role clarity, which improve team processes and outcomes.[158]

---

[156] Eduardo Salas, Michael A. Rosen, C. Shawn Burke, and Gerald F. Goodwin, "The Wisdom of Collectives in Organizations: An Update of the Teamwork Competencies," in Eduardo Salas, Gerald F. Goodwin, and C. Shawn Burke, eds., *Team Effectiveness in Complex Organizations: Cross-Disciplinary Perspectives and Approaches*, New York: Routledge, 2008, pp. 73–114.

[157] We did not find meta-analyses or reviews of general interpersonal skills training or training in conflict management that are relevant for the SOF population; these reviews tend to focus on children and individuals with psychiatric problems or developmental disabilities. We discuss training for Communication skills in a later section in this chapter.

[158] For example, see Christina Lacerenza, Shannon L. Marlow, Scott I. Tannenbaum, and Eduardo Salas, "Team Development Interventions: Evidence-Based Approaches for Improving Teamwork," *American Psychologist*, Vol. 73, No. 4, 2018, pp. 517–531.

Table 7.1. Teamwork Facets and Definitions

Teamwork	Working Productively with Others and Adapting Appropriately to Different Roles and Situations
Roles	• Performing assigned roles and responsibilities contributing to task achievement • Assisting others to perform their tasks (e.g., verbal encouragement, guidance, or behavioral assistance)
Monitoring	• Checking others' physical and mental states (e.g., hydration, individual gear) and understanding of roles • Tracking resources (e.g., team equipment, time)
Confidence Building	• Creating a supportive environment that expresses interest in others' ideas and feedback • Establishing the belief and expectations that high levels of performance are possible

Training for Specific Teamwork Competencies

In this section, we introduce and define four critical Teamwork competencies: (1) team orientation, (2) emotional intelligence, (3) empathy, and (4) leadership. Although these competencies represent different areas to target, some competencies are more amenable to change.

Team Orientation

O'Shea et al. discuss team orientation as a dispositional trait that can affect teamwork.[159] Driskell and Salas found that in contrast to team members who are more collectively oriented, members who are more egocentric tend to ignore or reject others' contributions.[160] Subsequent work suggests that more-team oriented individuals demonstrate higher levels of affiliation, or preferences to work with others rather than alone, and lower levels of dominance/rigidity, which reflects willingness to accept others' input.[161] In addition, it has been posited that these facets of team orientation are related to Big Five personality traits, such as agreeableness and extraversion, respectively.[162] Evidence shows that personality can change over a person's life span, often in response to environmental factors, such as opportunities to serve in new social and work roles, but such changes are incremental and may take years before they are apparent.[163] Thus, while serving in SOF might be associated with changes in personality traits, training interventions focused on personality change are not likely to be fruitful.

---

[159] Patrick G. O'Shea, James E. Driskell, Gerald F. Goodwin, Michelle L. Zbylut, and Stephanie M. Weiss, *Development of a Conditional Reasoning Measure of Team Orientation*, Washington, D.C.: American Institutes for Research, 2004.

[160] James E. Driskell and Eduardo Salas, "Collective Behavior and Team Performance," *Human Factors*, Vol. 34, No. 3, 1992; O'Shea et al., 2004.

[161] James E. Driskell et al., *Team Situational Awareness: A Research Model and Preliminary Analyses*. Orlando, FL: Naval Air Warfare Training Systems Division, 1997.

[162] O'Shea et al., 2004.

[163] Brent W. Roberts, Kate E. Walton, and Wolfgang Viechtbauer, "Patterns of Mean-Level Change in Personality Traits Across the Life Course: A Meta-Analysis of Longitudinal Studies," *Psychological Bulletin*, Vol. 132, No. 1, 2006.

Emotional Intelligence

Emotional intelligence (EI) is a controversial topic in the social sciences, with debates about its definition, how to measure it, and whether people can be trained to have it. One view is that EI is a cognitive ability allowing individuals to understand emotions in themselves and others. Another view, called the mixed model, is that EI consists of a collection of largely noncognitive skills, such as personality traits and social skills. Researchers have argued that EI is associated with a variety of interpersonal competencies in work contexts as well as with team performance. A variety of methods have been used in EI training, including lectures, discussion, practice, provision of feedback, and other approaches.

One meta-analysis found moderate, positive effects of EI training.[164] Results were similar for studies using ability-based and mixed models of EI and for studies that used pre-post designs or treatment and control groups. There were no effects related to the gender composition of groups, suggesting that EI training is effective for both women and men. Preliminary findings indicate that experiential methods involving feedback are more effective than are passive approaches such as lectures.[165]

Empathy

Empathy is an aspect of the Big Five trait of agreeableness. In the social sciences, there is a general consensus that there are three types of empathy: (1) cognitive empathy, which is an intellectual understanding of another person's emotions; (2) affective or emotional empathy, in which the individual experiences the same emotional reactions as another person; and (3) behavioral empathy, which is reflected in taking action in response to cognitive or affective empathy. High levels of empathy are associated with positive interpersonal outcomes (better relationships and prosocial behavior, such as voluntarily acting in ways to help others), and low levels of empathy are associated with negative interpersonal outcomes such as bullying and sexual misconduct.[166] Training can target all three types of empathy; it can consist of experiential approaches, lectures, demonstrations, and/or practice.

In their meta-analysis of randomized controlled trials, van Berkhout and Malouff found that that empathy training has moderate-to-large benefits and is effective for all three types of empathy.[167] The results also show several moderating factors. For example, training has stronger positive effects among college students and health professionals compared with other adults or youth. Similarly, methods using four principles of behavioral skills training—instruction, modeling, practice, and feedback—are more effective than methods using fewer principles; and objective measures of outcomes (e.g., tests to interpret others' emotions) are more effective than self-reported measures.

---

[164] Victoria Mattingly and Kurt Kraiger, "Can Emotional Intelligence Be Trained? A Meta-Analytical Investigation," *Human Resource Management Review*, Vol. 29, No. 2, 2019, 140–155.

[165] These findings are suggestive rather than conclusive because many studies did not provide sufficient information about training methods, and the sample of studies in the analysis was small.

[166] Emily Teding van Berkhout and John M. Malouff, "The Efficacy of Empathy Training: A Meta-Analysis of Randomized Controlled Trials," *Journal of Counseling Psychology*, Vol. 63, No. 1, 2016, pp 32–41.

[167] Teding van Berkhout and Malouff, 2016.

Leadership

Lacarenza et al. conducted a meta-analysis examining effectiveness of leadership training and development programs on Kirkpatrick's four outcomes: reactions, learning, transfer, and results.[168] Results show medium-to large, positive effects on all four outcomes, although effects depend on different aspects of training, such as content, delivery methods, and type of outcome. Given the importance of job behavior, we focus on findings for transfer, or the extent to which trainees apply knowledge and skills learned in training to behaviors on the job. Of particular relevance to AFSPECWAR, stronger training transfer was found for (1) skill-based outcomes compared with affective outcomes; (2) low-level leaders compared with midlevel and senior leaders; (3) programs using multiple methods (information-based, practice, and demonstration); (4) programs that included feedback; (5) programs based on needs analysis; (6) programs consisting of multiple sessions or using a "distributed" training approach (rather than a single session, called "massed" training); and (7) longer programs.[169] In addition, programs had positive effects for all types of competencies (e.g., business, interpersonal, and intrapersonal), although training had stronger effects for business skills than for other content areas.

Evidenced-Based Practices for Teamwork Training

Several meta-analyses and reviews have analyzed the effectiveness of different method of team training. Some results refer to specific types of outcomes, including affective outcomes (e.g., attitudes such as trust and cohesion), cognitive outcomes (e.g., knowledge of the team's task, situational awareness, and accurate team mental models), and team processes (e.g., communication and coordination).[170]

Team Interventions in General

Two meta-analyses have examined the effectiveness of team training in general. Salas et al. examined the effects of team training interventions on affective (e.g., trust, confidence in members' abilities, and attitudes about the effectiveness of team processes), cognitive (declarative team knowledge), and process (e.g., behaviors including communication, coordination, decisionmaking, and self-correction) outcomes.[171] They also analyzed the effects that moderator variables, such as training content and team characteristics, have on these outcomes. Overall, they found that team training has

---

[168] Christina N. Lacerenza, Denise L. Reyes, Shannon L. Marlow, Dana L. Joseph, and Eduardo Salas, "Leadership Training Design, Delivery, and Implementation: A Meta-Analysis," *Journal of Applied Psychology*, Vol. 102, No. 12, 2017; Donald L. Kirkpatrick, "Evaluating Training Programs: The Four Levels," San Francisco: Berrett-Koehler, 1994.

[169] Lacerenza et al., 2017.

[170] Shared mental models refer to team members "being on the same page" with respect to their cognitive understanding of the team's task, members' roles, and other aspects of team functioning.

[171] Eduardo Salas, Deborah DiazGranados, Cameron Klein, C. Shawn Burke, Kevin C. Stagl, Gerald F. Goodwin, and Stanley M. Halpin, "Does Team Training Improve Team Performance? A Meta-Analysis," *Human Factors*, Vol. 50, No. 6, 2008, pp. 903–933.

moderate positive effects on team outcomes.[172] Results show moderate positive effects for affective, cognitive, and process outcomes, although results were tentative for process outcomes. Training content is effective for all outcomes, whether the content focuses on taskwork, teamwork, or a combination of the two.

McEwan et al. analyzed four types of teamwork training or interventions: didactic approaches (e.g., classroom training); workshops (e.g., discussions about team goals); simulations (e.g., practicing skills in an environment similar to the actual operating environment, such as flight simulators); and in situ team reviews (e.g., after-action reviews).[173] In comparison with control groups, groups that received team training show moderate-to-large benefits for teamwork behavior, and all four types of training enhance team performance. Other findings include the following:

- More interactive approaches (i.e., workshops, simulations, and team reviews) enhance teamwork behaviors, but didactic approaches do not.
- Effects on team behaviors are consistent across most contexts (e.g., military teams, academic teams) and for both intact and newly formed teams, although the effect is stronger for new teams.
- Effects are positive whether team behaviors are measured by third-party observers or team members self-report, although the effect is larger for third-party measures.

Review findings by Goodwin et al. show that military team training can improve teamwork.[174] For example, they report on results for Team Dimensional Training (TDT), which is a program in the Navy's research program, Tactical Decision Making Under Stress. TDT, which uses guided team self-correction—a specific approach to debriefings based on an expert model of teamwork—has positive impacts on team coordination, adaptation, and other outcomes, such as accurate mental models, team leadership, and supporting behaviors.[175]

Team Building

Klein et al. define team building as "a class of formal and informal team-level interventions that focus on improving social relations and clarifying roles, as well as solving task and interpersonal problems that affect team functioning."[176] Team building is typically conducted in settings that do not mirror the actual work environment. It may involve a range of activities, such as outdoor experiences,

---

[172] Moderator variables affect the association of independent and dependent variables—that is, the relationship of the independent and dependent variables *depends* on the moderator variable. For example, the type of training can be a moderator variable because the effectiveness of training depends on whether the type of training is more or less interactive (see Desmond McEwan, Geralyn R. Ruissen, Mark A. Eys, Bruno D. Zumbo, and Mark R. Beauchamp, "The Effectiveness of Teamwork Training on Teamwork Behaviors and Team Performance: A Systematic Review and Meta-Analysis of Controlled Interventions," *PloS One*, Vol. 12, No. 1, 2017, e0169604).

[173] McEwan et al., 2017.

[174] Gerald F. Goodwin, Nikki Blacksmith, and Meredith R. Coats, "The Science of Teams in the Military: Contributions from over 60 Years of Research," *American Psychologist*, Vol. 73, No. 4, 2018, pp. 322–333.

[175] Kimberly A. Smith-Jentsch, Janis A. Cannon-Bowers, Scott I. Tannenbaum, and Eduardo Salas, "Guided Team Self-Correction: Impacts on Team Mental Models, Processes, and Effectiveness," *Small Group Research*, Vol. 39, No. 3, 2008.

[176] Cameron Klein, Deborah Diaz Granados, Eduardo Salas, Huy Le, C. Shawn Burke, Rebecca Lyons, and Gerald F. Goodwin, "Does Team Building Work?" *Small Group Research*, Vol. 40, No. 2, 2009, pp. 181–222.

discussions about group processes, and role-playing exercises, and can occur over a wide duration—from days to months.[177]

The results of a meta-analysis suggest that team building has moderate benefits for affective outcomes (e.g., trust) and process outcomes (e.g., coordination and communication) and negligible effects for cognitive outcomes (e.g., knowledge of team competencies).[178] Moderator analyses suggest that role clarification and goal-setting aspects of team building have moderate effects on team functioning, with smaller effects from aspects emphasizing interpersonal relations or problem solving. Effects also vary based on team size, with a moderate-to-large effect for large teams (more than ten members) and a small-to-moderate effect on small teams (fewer than five members) and medium-size (five to ten members) teams.

Debriefing

Studies of team reviews specifically (in this case, debriefs) can enhance such team processes as shared mental models about taskwork and teamwork. Teams that engage in debriefs perform better than those that do not, and well-conducted debriefs can improve effectiveness by as much as 25 percent. Allen et al. emphasize that several practices used to enhance debriefs are especially relevant to AFSPECWAR.[179] For example, acknowledge that there is ambiguity about team performance; use diagnostic debriefs (addressing how to improve); use trained facilitators; encourage all team members to participate; ensure psychological safety (so that team members feel comfortable in participating); reflect on group goals, methods, and processes; focus on few rather than many critical issues; and provide objective data.

Conclusions

Based on our review of the research, it is clear that some competencies, such as team leadership, EI, and empathy, can be trained, and there are a variety of effective training methods. To get the most out of Teamwork training, AFSPECWAR should apply the following recommendations:

- Incorporate training on EI, empathy, and team leadership into AFSPECWAR training.
- Use multiple evidenced-based instructional methods, including information-based approaches (e.g., lectures, reading assignments) and practical exercises, accompanied by diagnostic feedback.
- Use trained facilitators during training.
- Use theory and evidence to identify conditions that facilitate the transfer of training, including training content, characteristics of trainees, and characteristics of the work environment such as organizational supports.

[177] In contrast, team training is defined as a formal, systematic approach to build specific skills; it typically includes opportunities for practice and is conducted in the organizational setting (Klein et al., 2009).

[178] Klein et al., 2009.

[179] Joseph A. Allen, Roni Reiter-Palmon, John Crowe, and Cliff Scott, "Debriefs: Teams Learning from Doing in Context," *American Psychologist*, Vol. 73, No. 4, 2018, pp. 504–516.

- Evaluate training effectiveness using multiple measures, including objective and third-party measures of team outcomes.

Communication

Attribute Definition and Background

As evident in the previous section, Communication is closely related to other attributes such as Teamwork and Problem Solving. Because of these interrelationships, Krueger et al. "consider[ed] communication to be the quality of communication rather than the simple act of exchanging information."[180] Table 7.2 presents the definitions and facets of Communication.

Table 7.2. Communication Facets and Definitions

Communication	Expressing Thoughts and Ideas to Convey Information and Influence Others
Message Organization	• Structuring information logically, succinctly, and with purpose • Incorporating examples, evidence, or materials (e.g., visual aids) to explain and reinforce the message • Using correct grammar and appropriate terms to enhance audience understanding
Message Delivery	• Using appropriate variety in volume, pace, and tone to heighten and maintain interest • Using appropriate physical behaviors to support a verbal message (e.g., posture, hand gestures)
Active Listening	• Demonstrating listening and receipt of a message (e.g., taking notes, asking questions, making eye contact) • Summarizing the main points of a message

Communication is emphasized by the USAF as written in its handbook *The Tongue and Quill*:

> The men and women of the United States Air Force must communicate clearly and effectively to carry out our missions. Although we live in an era of rapid personal and mass communication that was barely imagined just a few years ago, our Air Force still requires face-to-face briefings, background papers, and staff packages to keep the mission moving forward.[181]

This handbook is an extensive resource covering the basics of effective communication and delves deeply into writing, speaking, and listening. The USAF also teaches communication in its professional military education instruction, giving students an opportunity to learn the principles, practice, and receive feedback.

[180] Krueger et al., 2022, p. 57.

[181] Air Force Handbook (AFH) 33-337, *The Tongue and Quill*, Washington, D.C.: Secretary of the Air Force, May 27, 2015, Incorporating Change 1, November 19, 2015, p. i.

Evidence-Based Practices for Developing Communication Skills

Communication skills are highly valued by employers, which in turn is reflected in student curriculum in two- and four-year colleges and universities and many types of leadership training. Surveys of the basic communication course in colleges and universities have been ongoing for more than four decades. Communication instruction contributes to the professional and personal success of students, given the importance employers place on communication skills in their selection processes.[182] Critical thinking is an area of emphasis because it is a cornerstone of developing and presenting oral communication. Communication courses also place a lot of emphasis on public speaking.[183]

Oral Expression

Costigan and Brink talk about the positive relationship between oral expression, listening, and influence, which is critical for workplace success. The authors reflect on the position, posited by Dale Carnegie in 1936, that "influence mainly boils down to the way in which we communicate with others" and is relevant not only to hierarchical relationships but also among peers.[184] Moreover, they examined how oral communication skills can be developed using leaderless group discussions and an oral presentation followed by a question-and-answer session—two approaches that are similar to interactions in the workplace (e.g., group problem solving and a mission briefing). The unpredictability of the question-and-answer session in particular stresses listening and oral presentation skills. The authors found that both techniques were effective in developing oral presentation skills and that students responded favorably to the training with the intention of taking the learned skills back to their jobs. In turn, the study also found that communication skills were correlated with interpersonal influence, an important goal for many interactions.

Much of the literature centers on interactions between medical professionals and patients. Although less directly related to the SW community, these studies point to the potential effectiveness of Communication training. For example, Kerr et al. conducted a systematic review exploring whether nurses' communication skills improved after training.[185] Their findings showed that training could improve nurses' communication skills. The systematic review of Communication training courses in oncology conducted by Barth and Lannen came to a similar conclusion.[186] The training courses included in this review required active participation of the students in such activities as role-play and lasted at least six hours. The study concluded that this type of training showed promise in changing communication behavior and attitudes. It also concluded that a continuous learning approach to maintain and refresh skills could be valuable.

---

[182] Sherwyn P. Morreaele, David W. Worley, and Barbara Hugenberg, "The Basic Communication Course at Two- and Four-Year U.S. Colleges and Universities: Study VIII—The 40th Anniversary," *Communication Education*, Vol. 59, No. 4, 2010, pp. 405–430.

[183] Morreaele et al., 2010.

[184] Robert D. Costigan and Kyle E. Brink, "Developing Listening and Oral Expression Skills: Pillars of Influential Oral Communication," *Journal of Management Education*, Vol. 44, No. 2, 2020, p. 130.

[185] Debra Kerr, Joan Ostaszkiewicz, Trisha Dunning, and Peter Martin, "The Effectiveness of Training Interventions on Nurses' Communication Skills: A Systematic Review," *Nurse Education Today*, Vol. 89, 2020, pp. 1–13.

[186] J. Barth and P. Lannen, "Efficacy of Communication Skills Training Courses in Oncology: A Systematic Review and Meta-Analysis," *Annals of Oncology*, Vol. 22, 2011, pp. 1030–1040.

Active Listening Skills

Active listening is critical to interpersonal relationships. Active listening involves not only verbal and nonverbal components but also a variety of advanced skills that include reflecting feelings, reflecting meaning (by repeating back important information), interpretation, immediacy (real time), caring, and self-disclosure. Leaders who are good listeners are better able to build trust with subordinates, exert influence over the workforce, and accomplish their mission.[187]

In a study to evaluate the benefits of teaching active listening skills, Bjornestad and colleagues developed, implemented, and evaluated an active listening curriculum taught to Reserve Officer Training Corps (ROTC) cadets at two universities. [188] The program incorporated learning and practice components using "PowerPoints, vignettes, and experiential exercises. The vignettes and role-play activities included scenarios that occur in military settings."[189] The focus of the curriculum was on experiential learning, which was "identified as the most useful instructional method of teaching active listening skills."[190]

To determine whether the curriculum was effective, the researchers compared scores on the Active Empathetic Listening Scale between a group of 37 ROTC cadets who were enrolled in the new curriculum and a control group of 27 cadets who were not. The results indicated a significant improvement in scores over time for cadets who received the instruction compared with the cadets who did not. Moreover, the study included "preliminary evidence of the importance of providing opportunities for the learning, application, and practice of active listening skills."[191]

Importance of Practice and Feedback

A common thread through much of the literature is the need to go beyond learning and employ instruction techniques that include student interaction and feedback. As with many nontechnical attributes, feedback and the right type of feedback plays an important role in developing communication skills.

In *The Tongue and Quill*, the USAF talks about "fighting for feedback" and the contribution that feedback can make to improving communication.[192] A 1996 assessment of communication skills curricula of Air Force professional military education programs reached a similar conclusion about the importance of feedback, recommending that the Air Force incorporate feedback into all training curriculum.[193] This assessment also emphasized that Communication skills training is more effective if it is context-based, which helps students apply what they learn on the job, and that exercises such as

---

187 Andrea Bjornestad, Seth Olson, and Lee Weidauer, "Improving Officer-Soldier Communications Through Active Listening Skills Training with Army ROTC Cadets," *Military Psychology*, Vol. 33, No. 3, 2021, pp. 205–215.

188 Bjornestad et al., 2021.

189 Bjornestad et al., 2021, p. 207

190 Bjornestad et al., 2021.

191 Bjornestad et al., 2021, p. 213.

192 AFH, 2015, p. 104. The manual also provides information on where to get feedback, what kind of feedback is useful, and even how to give feedback.

193 Frank E. Jones, Yueh-Jyh Jong, Mark H. Jordan, Edwin T. Parks, Abigail Gray, Deanna A. Paulk, Lawrence, G. Ruggiero, Stanley W. Shrader, Leneva J. Schroeder, and Rex Jordan, *Assessing the Communication Skills Curricula of Air Force Professional Military Education Programs*, research paper, ACSC/DR/078/96-04, April 1996.

role-playing should relate to the type of work and the circumstances under which it is conducted in a student's unit.

In her study of crisis communication skills, Foote asserted that the best way to learn how to communicate during a crisis environment is through practice.[194] Foote developed an MBA elective course that used a student-centered learning approach involving interactive media tools, such as online videos and group role-plays, and that required students to develop crisis action plans and deliver a press conference during a simulated crisis situation. Her course design provided opportunities to practice and receive feedback, which are integral to developing communication skills. In short, Foote concludes that learning is not enough; students build capabilities through practice, problem solving, and feedback.

Morin and Latham explored the effect of other post-training techniques, mental practice (rehearsal of a task), and goal-setting, which are aimed at developing self-efficacy and can help students transfer the communication skills they are learning into a work setting.[195] Prior to the study, supervisors had participated in a one-day training on interpersonal communication skills but had not mastered applying these skills on the job. A one-day refresher program was conducted that included a lecture, group discussion, video presentations, and role-play. The results of the study showed that behavior on the job improved for supervisors who engaged in mental practice (or mental practice in combination with goal-setting). In other words, mental practice helped students symbolically achieve mastery, which they were then able to apply in the workplace.

Conclusions

This review points to the overall value of Communication training—in building leaders, in developing effective team interactions, and in building up capabilities to be influential in the workplace. But to be effective, learning theory must be combined or followed with student-centered learning opportunities that provide opportunities to practice and receive feedback, especially within the context of performing their job duties.

---

[194] Laura M. Foote, "Honing Crisis Communication Skills: Using Interactive Media and Student-Centered Learning to Develop Agile Leaders," *Journal of Management Education*, Vol. 37, No. 1, 2013, pp. 79–114.

[195] Lucie Morin and Gary Latham, "The Effect of Mental Practice and Goal Setting as a Transfer of Training Intervention on Supervisors' Self-Efficacy and Communication Skills: An Exploratory Study," *Applied Psychology*, Vol. 49, No. 3, 2000, pp. 566–578.

Chapter 8

Conclusions and Recommendations

AFSPECWAR attributes are the foundation for A&S, but the integration of these attributes in follow-on training courses is uneven and not systematically integrated into course objectives and events. This should not be interpreted to mean that AFSPECWAR does not develop nontechnical attributes. On the contrary, our content analysis of training documents demonstrated that nontechnical attributes can be aligned to training objectives and events. Furthermore, both instructor and student survey responses indicated that training courses generally emphasized each of the attributes. However, this development is not deliberate and occurs primarily through exposure to technical training instruction.

The majority of students in AFSPECWAR training are performing well, but gaps were identified in four attributes:

- Teamwork (e.g., interpersonal skills)
- Drive
- Trainability
- Stress Tolerance.

Although the underlying causes of these gaps cannot be determined without more extensive follow-up analysis with individual students, multiple strategies should be considered to bolster different parts of the talent management system. Effective interventions should focus both on the individual (e.g., goal-setting) and the training environment (e.g., establishing and maintaining a strong culture by encouraging prompt and constructive feedback and providing appropriate reinforcements for performance improvements). Furthermore, research suggests that attribute development should be increasingly job-specific as students progress through the training pipeline.

In this chapter, we expand on these findings and recommend strategies to bolster development of attributes in the training aspects of the talent management system.

Recommendations

Guided by this needs analysis and a targeted review of the research literature, we developed a roadmap with seven broad recommendations that could be applied by SWTW and DAF to develop nontechnical attributes (see Figure 8.1). These recommendations are aligned with a roadmap that can build off the training needs analysis that we conducted. The recommendations also align with broader efforts in the U.S. educational system to design education systems around clearly specified outcomes for students (i.e., what they should be able to do after they complete their education). This concept of

Figure 8.1. Recommendation Roadmap for Nontechnical Attribute Development

1	2	3	4	5	6	7
Conduct periodic needs analyses	Align attributes to course objectives/ events	Create conditions for positive training transfer	Determine appropriate training strategies and interventions	Incorporate goal-setting for performance improvement	Provide timely, diagnostic feedback	Conduct formal program evaluations

outcomes-based education is being applied to military settings as well, notably Joint Professional Military Education.[196]

1. Conduct Periodic Needs Analyses

A needs analysis documents current training practices and determines potential gaps in attribute development. This analysis is particularly important for organizations that have finite resources to focus attention on areas with the greatest deficiencies. We completed a needs analysis as part of this project using a combination of performance evaluations and discussions with course instructors, leadership, and students. When possible, organizations should use multiple approaches and different sources of input (see Appendix A for other approaches). Although convergence among data sources is desired, discrepancies among sources can also provide insights (e.g., lack of knowledge about performance standards, limited opportunities to observe). Ideally, a needs analysis should be conducted across the full pipeline every one to three years to evaluate the effectiveness of different interventions and to address any new or emerging issues.

Figure 8.2 provides recommended sources of information for each attribute.[197] Specifically, we map each attribute (rows) against different sources of information about training needs (columns). Sources designed as primary (P in the cell) are those that should provide the main source of information about the attribute, whereas other sources may be secondary (S in the cell). For example, students need to provide some insight into their own sense of Drive, hence the "P" for "self" in the Drive row, but an operational psychologist could provide an additional (secondary) source of information about the student's Drive. In contrast, Physical Fitness is more accurately measured through tests and other assessments and should not rely on self-assessment, which is why "self" is

[196] Paul W. Mayberry, Charles A. Goldman, Kimberly Jackson, Eric Hastings, Hannah Acheson-Field, and Anthony Lawrence, *Making the Grade: Integration of Joint Professional Military Education and Talent Management in Developing Joint Officers*, Santa Monica, Calif.: RAND, RR-A473-1, 2021.

[197] Depending on the specific attribute, different sources of information can be used to evaluate how well students are performing on a specific attribute. For example, peers may be a useful source of information about other students' interpersonal attributes, including Teamwork and Communication. Some attributes such as Trainability and Physical Fitness can be evaluated using formal assessments. Specialized expertise may also be required to provide more accurate diagnostic information. For example, operational psychologists should be used to support evaluations of Stress Tolerance, and physical therapists and strength and conditioning coaches should be used to support evaluations of Physical Fitness.

Figure 8.2. Periodic Needs Analysis Using Multiple Sources

	Self	Peers	Instructors	Other SMEs	Operational Psychs	Athletic Trainers / Physical Therapists	FMPs	Objective Measure / Test
Trainability	P	P	P	S	P			P
Teamwork		P	P		S		P	
Stress Tolerance	P	S	P		P		P	
Problem Solving		S	P	P	S		P	S
Physical Fitness		S	S			P	P	P
Integrity	P	P	P		S			
Drive	P	P	P		S		S	S
Communication		P	P	P	S		P	

NOTE: P = primary source; S = secondary source.

neither a primary nor a secondary source for Physical Fitness. Figure 8.2 can therefore be considered a framework for selecting sources of information for a needs analysis of AFSPECWAR attributes.

2. Align Attributes to Course Objectives/Events

The feedback from course instructors and students consistently indicated that SWTW attributes are emphasized, albeit unevenly across courses, but not deliberately developed. To make attribute development more explicit, we recommend identifying which course objectives and events provide opportunities for development. Although not all objectives or events need to align to a specific attribute, linkages should be explicit and documented to focus the attention of instructors and students on developmental and feedback opportunities.

Use Alignment to Objectives and Events to Guide Attribute Evaluations

For most linkages of attributes to objectives and events, feedback should be encouraged and reinforced but not required. However, some events could present multiple opportunities to demonstrate one or more attributes. These events should be considered as sources for collecting formal evaluation data and for tracking individual-level changes over time (e.g., mission brief during FMP to evaluate Communication). These evaluations should be conducted periodically and as soon after the event has concluded to reinforce the importance and accuracy of evaluations. When determining how frequently to conduct evaluations, it is important to consider whether students have had enough time to improve and whether evaluations would interfere with other priorities (e.g., learning technical tasks).

Review and Refine Our Alignment of Attributes to Course Objectives and Events

Our alignment of attributes to objectives and events should be considered preliminary and should be reviewed (i.e., confirmed or updated as necessary) by course developers and instructors. More specifically, attributes and relevant facets should be documented in the course documents (e.g., plans of instruction [POIs]) to ensure increased awareness and opportunities to reinforce specific attributes.

Additionally, course designers may want to consider modifications to certain events if changes improve elicitation of attribute-relevant behaviors but do not interfere with the execution of the primary lesson plan.

3. Create Conditions for Positive Training Transfer

Learning and understanding what to do does not always result in performance improvement in the target environment (i.e., training transfer). There are several practices that can be followed to increase the transfer of SWTW training.

Invest in the Instructor Qualification Course

IQC provides the foundation for preparing instructors on how to deliver curricula. Consequently, instructors are the primary sources of feedback to students on how well they are acquiring technical knowledge and skills. Instructors are also well positioned to observe and provide feedback on nontechnical attributes. Although IQC has a limited amount of time to develop instructors, a key element of training should be devoted to training instructors on how to provide effective feedback and when to step back and allow students to self-correct mistakes. Discussions with students indicated that feedback from instructors was not always consistent or constructive.[198]

We also recommend that SWTW review processes and criteria for selecting instructors. Discussions with leadership indicated that interviews were used at one time to prescreen instructors, but that practice was abandoned because of resource demands and the limited number of staff to conduct the interviews. Establishing additional prescreening could help socialize new instructors and promote buy-in to attribute-centered development. Some options to consider include using a situational judgment test, a brief structured interview, and supervisor evaluations of past performance and potential as an instructor.

Reward Improvements Not Just Outcomes, Which Helps to Create a Learning Culture

Technical training incorporates multiple measures (e.g., progress checks, exams, and milestone events) that recognize and reward students for exceeding standards. However, standards for nontechnical attributes are generally undefined. Therefore, nontechnical attributes might need to be reinforced using a different approach. Specifically, the focus should be on rewarding students' efforts and improvement on attribute-relevant behaviors. If feedback emphasizes what students should start, stop, and continue doing, then recognition and rewards should be used to reinforce those behaviors. Focusing on improvement will require a more individualized approach compared with traditional AETC training practices that rely heavily on the implementation of common standards for all

---

[198] Students also have an opportunity to provide feedback about instructors by scanning a QR code that is posted around classrooms. However, discussions suggested that the scrutiny on instructors reflects an environment of distrust and results in instructors having to dedicate more time to paperwork to justify decisions they make about students, especially when student performance is substandard.

students in an AFSC. That is, SWTW will need to establish baselines for individual students and then monitor and document progress by comparing against initial baselines.

Establish Relevance of the Training by Tying It to Real-Life Operational Experiences

Gaining the commitment of students to develop nontechnical attributes may be difficult, especially if training lacks perceived relevance to the jobs and missions they will perform. These risks may be particularly high at the beginning of the pipeline when students have limited knowledge of the jobs and mission environments. To increase relevance of nontechnical training, instructors should catalog operational experiences that can be used to directly reinforce the importance of specific attributes and facets. For example, how you handle your weapon can establish credibility, which is especially important when operating with unfamiliar personnel from the sister services. That is, does your weapon look clean? Do you hold your weapon tactically? Is your weapon within arm's reach? These examples highlight how paying attention to details (the facet of Reliability within Drive) can affect the trust that others place in you.

Establish a Climate for Providing Peer Feedback

Discussions with students indicated a desire to know the source of the feedback they receive, which is preferred over anonymous, electronic peer feedback that lacks context and specificity for making improvements. There appears to be a range of practices in how peer feedback is executed. For example, some students indicated that hot washes are used to provide face-to-face feedback, but others said that electronic ratings and printouts with feedback are used. Students who have participated in a hot wash say that these sessions can be useful in resolving discrepancies among those providing feedback. For example, a student may receive negative feedback from a peer, but that peer did not have all the facts or the full context. In this case, other peers can voice their perspectives at the same time, sharing additional information to help improve the accuracy of feedback. Although this practice increases the accountability of those giving the feedback, sessions can become hostile if not monitored by a trained facilitator. Furthermore, some feedback may be best handled in one-on-one sessions, especially if the content of the feedback is sensitive.

We recommend continuing the practice of providing peer raters with training on how to evaluate and provide behavioral feedback. This training may need to be augmented with practice opportunities and feedback on how well they are providing feedback. We also encourage practices that use incentives or accountability mechanisms (e.g., having to justify ratings/feedback to the lead instructor) to promote quality and accuracy of feedback. If using a hot wash to provide feedback, ensure raters have sufficient time to collect and document their thoughts. The documentation should be shared with students because retaining information during feedback sessions can be difficult.

Finally, it is important to ensure that every student is receiving feedback from multiple peers and that the number of peers providing feedback to any single student is similar to the number of raters for other peers. Operational psychologists may need to monitor the feedback providers and documentation to reduce imbalances.

4. Determine Appropriate Training Strategies and Interventions

The results of the training needs analysis should be used to determine the timing and appropriate strategy for attribute development. Although we focus primarily on training courses, there are opportunities for nontechnical attribute development when students are out of training or awaiting training (e.g., recovering from an injury in SW development). Additionally, there are several factors that should be factored into nontechnical training design, including the target attribute, stage of training, characteristics of the student (e.g., motivation, self-efficacy), and resource requirements (e.g., instructors, length of training, technology). Furthermore, the type of developmental strategies and experiences that are effective before students have graduated the pipeline may not be sufficient to promote attribute development as operators mature in their careers. Indeed, other forms of development such as targeted assignments that require solving more complex or ill-defined problems can be used with more senior operators.

Some Interpersonal Skills Gaps Are Best Addressed Early in the Training Pipeline

Several of the gaps identified by the training needs analysis suggest that SWTW should implement additional training and steps to strengthen and shape the culture early. For example, ineffective Teamwork behaviors indicated by constant complaining, negative mindsets, and not helping others are important to correct in BMT or SWCC. Increasing the presence of active-duty members in SWCC, providing interpersonal skills training, and giving prompt feedback should be considered. It is important to set the stage early by setting clear expectations and holding students accountable for desired behaviors.

Increase the Specificity of Nontechnical Attribute Development as Students Get Closer to the Job

As skills become more complex and as students move through the pipeline, training strategies should more closely approximate the operational environment and corresponding demands. Physical Fitness training should also shift to a more narrowed focus on developing physical skills needed to perform specific occupational tasks in specific environments (e.g., rucking through mountains). Increasing the specificity of nontechnical attribute development aligns with current technical training practices (e.g., crawl, walk, run).

Some attributes are best developed on a foundation of content knowledge and experience. For example, general Problem Solving frameworks can be introduced early in the training pipeline (e.g., emphasizing the importance of defining the problem first); however, resolving complex problems on-the-job often requires content expertise to fully understand the problem and generate appropriate solutions. Introducing complex frameworks too early can increase the cognitive load on students and ultimately interfere with learning core technical skills. Similarly, some training strategies (e.g., promoting independent learning) are more appropriate for students later in the pipeline when they have established a foundation of basic knowledge and skills.

Course instructors should also be trained to introduce and emphasize the importance of informal learning and principles of self-development that may be more important once students complete the formal training pipeline.[199] Course instructors could provide students "with tips and insights about effective [informal learning behaviors] ILBs (e.g., to become better at asking questions, listening actively, recognizing patterns, soliciting feedback, or observing experts)."[200]

Figure 8.3 provides a framework for developing attributes based on the stage of the training pipeline (columns). For a given attribute in a training stage, we offer suggestions on what the focus should be, based on the themes outlined above (e.g., attribute development becomes increasingly job-specific as students move through the pipeline). The content for a given attribute changes as students move across the phases. In some cases, we offer suggestions on specific instructional methods (e.g., team debriefs) that can be used. However, for most cases, we leave open the instructional approach to use, allowing for a blend of approaches (e.g., classroom instruction, modeling, simulations, practice) across phases. That said, student attributes (e.g., motivation, receptiveness to feedback) should be taken into consideration when selecting the type of instructional approach to use for a given attribute at a particular phase of training and development.

---

[199] Karin A. Orvis and Krista Langkamer Ratwani, "Leader Self-Development: A Contemporary Context for Leader Development Evaluation," *The Leadership Quarterly*, Vol. 21, No. 4, 2010, pp. 657–674.

[200] Christopher P. Cerasoli, "Antecedents and Outcomes of Informal Learning Behaviors: A Meta-Analysis," *Journal of Business and Psychology*, Vol. 33, 2018, p. 225.

Figure 8.3. Strategies for Progressive Attribute Development for AFSPECWAR

Attribute Development Should Be Increasingly Job-Specific →

	BMT/SWCC	IST	FTU/MAJCOM	Operational / OJT
Trainability	• Emphasize importance of feedback acceptance	• Refine meta-cognitive skills • Establish formal mentoring • Create conditions for positive training transfer	→	→
Teamwork	• Establish expectations • Focus on generic & transportable teamwork skills (e.g., planning, coordination, interpersonal skills) • Training to improve understanding others' perspectives to include instruction, modeling, practice, & feedback	• Recognize & identify team roles • Incorporate simulations & team reviews • Model & talk out effective & ineffective behaviors	• Conduct team debriefs using trained facilitators	• Conduct team debriefs using trained facilitators • Build context-specific competencies (e.g., backup behaviors) with intact teams
Stress Tolerance	• Introduce psychological skills • Initial practice	• Focus on mastering job skills and building confidence • Discuss and identify relevant stressors	• Apply psychological skills during simulations	→
Problem Solving	• Introduce general concepts	• Provide general problem-solving frameworks to support knowledge/skill acquisition	• Provide mission-specific frameworks from SMEs	→
Physical Fitness	• Build broad foundation across all abilities • Develop physical skills (e.g., water)	• Maintenance	→	• Focus on abilities/skills needed to perform specific mission tasks (e.g., grip strength)
Integrity	• Establish expectations	• Monitor & hold accountable	• Provide mission-specific ethical dilemmas	→
Drive	• Establish expectations • Teach goal-setting and self-assessment	• Monitor & hold accountable	→	→
Communication	• Introduce general concepts	• Model effective communication • Discuss ineffective communication • Provide several practice opportunities	→	→

Consider the Resource Requirements Needed to Bolster Nontechnical Attribute Development

Based on discussions with SWTW, apprentice courses are being lengthened to resolve concerns about instructor workload and burnout. Adding more responsibilities (e.g., curriculum, feedback sessions) will only add to the current workload. Therefore, SWTW should evaluate the additional resources needed to formalize attribute evaluation, curriculum design, and the structure of individualized feedback and follow-up sessions. Providing individualized feedback and setting learning goals is time-intensive and requires commitment beyond sending summary feedback through an automated system.

SWTW should also evaluate the potential benefits of investing in emerging technology solutions. For example, virtual reality can be an effective tool in delivering stress exposure training. Virtual reality could also be explored as a solution to providing additional practice opportunities for both technical and nontechnical training.

5. Incorporate Goal-Setting for Performance Improvement

Goal-setting is one of the most robust and well-researched interventions.[201] Although there is no "one" format, goal-setting interventions typically have two characteristics: (1) individuals establish specific but challenging goals that align with performing one or more tasks, and (2) individuals must commit to the goal. Both characteristics influence individuals' efforts to achieve their goal(s). These interventions are often accompanied by assessments of self-efficacy and feedback sessions.

Continue Current Practice of Emphasizing Behaviors That Students Should Start, Stop, and Continue Doing

A somewhat common practice in the AFSPECWAR community is framing feedback in the form of "start," "continue," or "stop" (e.g., an after-action review). In some training courses, this framing has been applied to student feedback with the goal of providing actionable comments. We encourage this practice to continue and emphasize that those providing feedback should be instructed to focus on behaviors (an observable action that can be verified by others) as opposed to judgments (evaluative statements about performance that can have varying interpretations).[202]

Establish Performance Improvement Goals Focused on Attribute-Relevant Behaviors

Achieving performance goals often takes more than one attempt. For example, it takes oral communication skills as well as domain knowledge (i.e., knowing which information to relay) to effectively communicate operationally relevant information over a radio. Establishing improvement

---

[201] Van Iddekinge et al., 2022.

[202] Behaviors will involve verbs such as "generates" or "assists," whereas judgments will involve adjectives such as "smart" or "solid."

goals will help students demonstrate progress toward the goal. These improvement goals should allow students to demonstrate attribute-relevant behaviors (e.g., performing radio communications in a training exercise to demonstrate Communication). In evaluations of student demonstrations of these behaviors, instructors can tie performing the attribute-relevant behaviors to improvements in the attribute itself, which will help students understand how their progress is tied to both an occupational task and attribute-relevant behavior performance goals.

Use Strategies to Improve Commitment to Improvement Goals (That Are Specific, Measurable, and Challenging yet Attainable)

Although setting goals is a critical first step, recent research finds that committing to goals has more impact on work effort.[203] To enhance commitment, goals must be

- specific—be as detailed as possible
- measurable—identify appropriate metrics
- challenging—ensure it is difficult yet realistic (e.g., stretch goals).

Using the radio communications example above, students could start by presenting simple operational information to peers and instructors. Instructors and peers can use a rating rubric to evaluate the communication accuracy (e.g., Did the student present the right information? How many errors in terminology were made?) and quality (e.g., Did the student clearly enunciate so others could hear what they said?). Follow-on events would involve increasingly complex information and more realistic settings (e.g., instead of presenting in front of a classroom, have the student use a radio in a training exercise). After each event, instructors (and peers) would provide feedback that would indicate communication improvements made across the successive events so that the students can understand how their attribute improvements are tied to the communication-related performance goal.

6. Provide Timely, Diagnostic Feedback

Assess Student Feedback Orientation and the Broader Feedback Environment

As summarized by Bell et. al., "What happens prior to and after training can greatly influence training effectiveness, so it is critical to take a systems perspective that goes beyond instructional design and trainee characteristics and considers the context within which training occurs."[204] Consequently, SWTW should focus on characteristics of the feedback and learning environment, including supervisor and peer support, how feedback is delivered, and opportunities to practice and apply new skills. Specifically, we encourage SWTW to use validated measure(s) of students' receptiveness and likelihood to use feedback and the broader feedback environment (e.g., credibility and usefulness of feedback provided, how feedback is delivered). The results from these assessments

[203] Van Iddekinge et al., 2022.

[204] Bradford S. Bell, Scott I. Tannenbaum, J. Kevin Ford, Raymond A. Noe, and Kurt Kraiger, "100 Years of Training and Development Research: What We Know and Where We Should Go," *Journal of Applied Psychology* (Centennial Special Issue), Vol. 102, No. 3, 2017, p. 315.

can used to improve the effectiveness of feedback. Examples of these assessments include the Feedback Environment Scale (FES) and the Feedback Orientation Scale (FOS).[205] The FES should be administered within specific courses and frequently enough to account for instructor rotations. The FOS could be conducted once per student at the beginning of SWCC and as follow-ups later in the pipeline to determine if there have been changes in how students perceive the value of feedback.

Formalize Who, When, What, and How Feedback Is Delivered

In general, we encourage informal, frequent, and transparent feedback. However, formal feedback collected through electronic systems is useful for monitoring progress against learning goals and for promoting a common understanding between students and feedback providers on strengths and areas for improvement. Discussions with students and instructors indicated that such formal feedback is time-consuming and not always beneficial. To increase the usefulness of formal feedback, we recommend the following strategies:

a. Allow sufficient time for feedback providers to meet individually with students to discuss the feedback, review strategies for improvement, and set performance improvement goals.
b. Use credible source(s) (i.e., expertise and trustworthy) to provide feedback (see Figure 8.2).
c. Give constructive feedback with explanations to help students differentiate effective from ineffective performance behaviors.
d. Focus on specific behaviors (e.g., the student proceeded to eat his MRE while everyone else helped to clean up and put away team gear); avoid statements about a person's character or personality (e.g., the student is selfish).
e. Model effective feedback, and provide feedback to feedback providers.

Determine Who "Owns" and Has Access to Attribute Evaluations and Feedback

Some USAF-wide platforms, such as MyVector, are designed to maintain rater anonymity and limit access to the ratings and feedback. That is, supervisors and others in the chain of command do not have access to individual-level results contained in MyVector. These practices are generally supported by research in commercial industry to ensure that the information is used only for developmental purposes and prohibits use for administrative decisions (e.g., promotions or assignments). Although the user can share results with others, there is limited accountability for addressing performance deficiencies. In line with the culture of SWTW, in most situations we encourage open and transparent feedback that is delivered promptly and in a supportive and constructive manner. However, certain types of feedback may require more sensitive handling. We encourage SWTW to determine what sources and types of feedback will be made available to peers, instructors, and leadership. These decisions should be codified to promote consistency across training courses.

---

[205] Lisa A. Steelman, Paul E. Levy, and Andrea F. Snell, "The Feedback Environment Scale: Construct Definition, Measurement, and Validation," *Educational and Psychological Measurement*, Vol. 64, No. 1, 2004, pp. 165–184; for the FOS, see Linderbaum and Levy, 2010.

7. Conduct Formal Program Evaluations to Align Talent Management Practices

Although many strategies to promote nontechnical attribute development are already integrated into SWTW training practices, specific courses, programs, or curricula may be needed to target specific attributes or facets (e.g., virtual reality for stress exposure training). For formal interventions, we recommend that SWTW conduct systematic evaluations to determine whether goals are being met. When possible, SWTW should use a research group (e.g., SWTW Research Flight) independent of program developers to minimize potential conflicts of interest. These evaluations could be used as part of a broader talent management feedback loop that incorporates outcome data such as washbacks, eliminations, and graduations (Figure 8.4).

Figure 8.4. Program Evaluation Feedback Loop

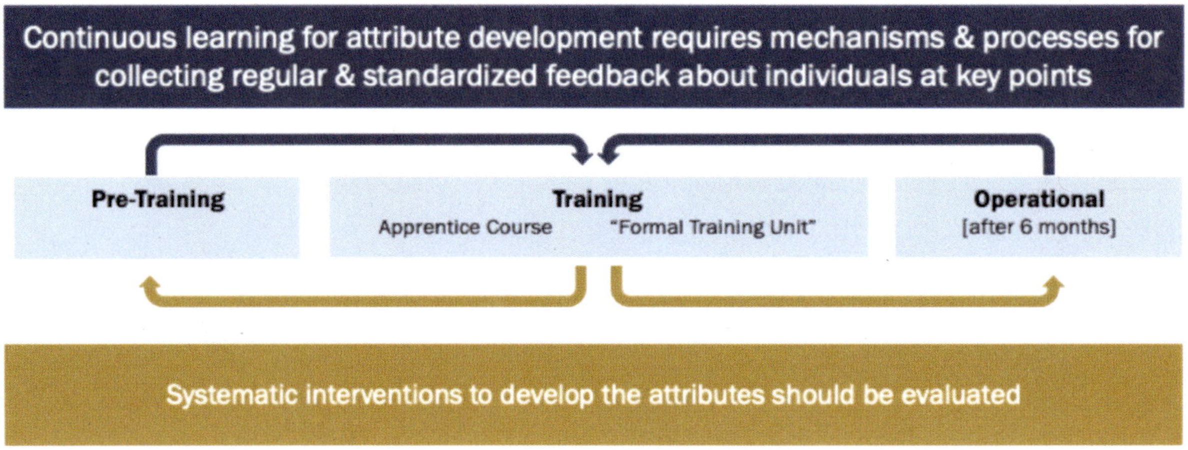

Implement Pilot Studies

Before implementing across all training courses, SWTW should consider conducting pilot studies with control groups to gain early insights on what features of the intervention may or may not be working. Alternatively, some students could be assigned to a "wait list" group that undergoes training after an initial group. Outcomes could be collected after the first group is trained but immediately before the wait list group begins training.

Consider Using Multiple Criteria to Evaluate Interventions

Nontechnical attribute development can be designed to affect multiple criteria, including affective, cognitive, behavioral, and performance (i.e., results) outcomes. SWTW already incorporates measures to evaluate student reactions (e.g., course evaluations) and multiple measures of learning for technical training. SWTW will need to continue these practices to evaluate nontechnical interventions and should consider additional measures to track changes in behaviors and performance. Measures should be taken at baseline, after training, and during post-training simulations or on-the-job performance. If possible, the same measurements should be taken at longer-term follow-ups (six months, one year, etc.) to assess the maintenance requirements of attribute development.

Prioritizing the Recommendations for AFSPECWAR

The recommendations provided in this chapter offer a general implementation roadmap that can be adopted by any functional community. There are more specific steps to facilitate implementation of the roadmap. We emphasize the importance of identifying the key stakeholders needed to successfully implement the attribute development roadmap. We further emphasize the following steps to address the attribute gaps (Drive, Teamwork, Trainability, Stress Tolerance). First, teach relevant frameworks as are applicable to current course progression in the pipeline. Second, allow students to practice and apply the frameworks. Third, use credible sources to provide feedback on what is working and areas for additional development. More specific strategies to address attribute gaps should be considered based on a thorough evaluation of current practices and student needs (see Figure 8.3).

Conclusion

Beyond the initial introduction to AFSPECWAR attributes (e.g., in SWCC), attributes are seldom discussed explicitly. However, attribute development is occurring primarily through exposure to technical training instruction and exercises, though our analyses suggest this exposure and development is uneven across training courses. AFSPECWAR can build on current practices by incorporating more deliberate attribute training, which may be needed to address gaps specific areas (e.g., interpersonal skills). This training should be progressive and incorporate prompt feedback from multiple credible sources. Implementing a comprehensive roadmap such as the one offered in this report will help to ensure future generations of SW operators are prepared to continue learning and adapting to an uncertain future.

Appendix A

The Science of Training and Development

When developing any type of training program, there are broad principles that should be considered. That is, these principles can be applied to technical training and nontechnical attribute development. These principles are crosscutting and relevant for developing multiple AFSPECWAR attributes. In this appendix, we introduce the concepts that are elaborated on in the chapters with more tailored discussions.

Needs Analysis

A needs analysis is often the first step in developing a personnel training program. The term refers to the systematic process of collecting and analyzing data to determine if personnel training would benefit an organization. A needs analysis seeks to systematically (1) categorize the goals of an organization, (2) identify knowledge and skill gaps within an organization that are relevant to those goals, and (3) determine how to best address those gaps through targeted personnel training.[206] Although conducting a thorough needs analysis may or may not affect the effectiveness of training,[207] understanding trainee needs helps to ensure that limited resources are allocated appropriately.

A needs analysis is a critical element of training design and implementation because the identification of training needs determines who and what will be trained.[208] A variety of methods can be used, including observations, reviews of existing organizational data, interviews, focus groups, surveys, questionnaires, performance appraisals, tests, assessment centers, and advisory committees.[209] It is generally recommended to use multiple methods during a needs analysis.[210]

---

[206] Irwin L. Goldstein, "Training in Work Organizations," *Annual Review of Psychology*, Vol. 31, No. 1, 1980, pp. 229–272.

[207] Winfred Arthur, Jr., Winston Bennett, Jr., Pamela S. Edens, and Suzanne T. Bell, "Effectiveness of Training in Organizations: A Meta-Analysis of Design and Evaluation Features," *Journal of Applied Psychology*, Vol. 88, No. 2, 2003, pp. 234–245.

[208] Paul J. Taylor and Michael P. O'Driscoll, "A New Integrated Framework for Training Needs Analysis," *Human Resource Management Journal*, Vol. 8, No. 2, 1998, pp. 1–29.

[209] Judith Brown, "Training Needs Assessment: A Must for Developing an Effective Training Program," *Public Personnel Management*, Vol. 31, No. 4, 2002, pp. 569–578.

[210] Brown, 2002.

Training Design

Training can be used to enhance individual capabilities, improving team effectiveness and helping organizations gain a competitive advantage.[211] Given the impact of training, it is important that both researchers and practitioners apply evidence-supported practices in the design of new training programs.[212]

Recent research has emphasized a theory of learner-centered training design.[213] Learning is an inferred process that can be detected through changes in an individual's knowledge, skills, attitudes, and behaviors, first during training and later on the job (or in the operational environment).[214] Essentially, learning can be thought of as "increased capacity to do the right thing at the right time."[215]

To facilitate learning, training uses various methods including technology-based training,[216] gaming,[217] experiential learning,[218] mentoring,[219] and coaching.[220] Training can also facilitate learning after instruction is finished with techniques that prompt self-directed learning,[221] informal learning,[222]

---

[211] Raymond A. Noe, Alena D. M. Clarke, and Howard J. Klein, "Learning in the Twenty-First-Century Workplace," *Annual Review of Organizational Psychology and Organizational Behavior*, Vol. 1, No. 1, 2014, pp. 245–275.

[212] Eduardo Salas and Janis A. Cannon-Bowers, "The Science of Training: A Decade," *Annual Review of Psychology*, Vol. 52, 2001, pp. 471–499.

[213] Kurt Kraiger and J. Kevin Ford, "The Science of Workplace Instruction: Learning and Development Applied to Work," *Annual Review of Organizational Psychology and Organizational Behavior*, Vol. 8, No. 1, 2021, pp. 45–721; Bell et al., 2017; Noe, Clarke, and Klein, 2014.

[214] Timothy T. Baldwin and J. Kevin Ford, "Transfer of Training: A Review and Directions for Future Research," *Personnel Psychology*, Vol. 41, No. 1,1988, pp. 63–105.

[215] Kurt and Ford, 2021.

[216] Wendy L. Bedwell and Eduardo Salas, "Computer-Based Training: Capitalizing on Lessons Learned," *International Journal of Training and Development*, Vol. 14, No. 3, 2010, pp. 239–249.

[217] Thomas M. Malaby, "Beyond Play: A New Approach to Games," *Games and Culture*, Vol. 2, No. 2, 2007, pp. 95–113; Traci Sitzmann, "A Meta-Analytic Examination of the Instructional Effectiveness of Computer-Based Simulation Games," *Personnel Psychology*, Vol. 64, No. 2, 2011, pp. 489–528.

[218] David A. Kolb, *Experiential Learning: Experience as the Source of Learning and Development*, Upper Saddle River, N.J.: FT Press, 2014.

[219] Kurt Kraiger, Lisa M. Finkelstein, and Lebena S. Varghese, "Enacting Effective Mentoring Behaviors: Development and Initial Investigation of the Cuboid of Mentoring," *Journal of Business and Psychology*, Vol. 34, No. 4, 2019, pp. 403–424.

[220] Rebecca J. Jones, Stephen A. Woods, and Yves R. F. Guillaume, "The Effectiveness of Workplace Coaching: A Meta-Analysis of Learning and Performance Outcomes from Coaching," *Journal of Occupational and Organizational Psychology*, Vol. 89, No. 2, 2016, pp. 249–277.

[221] Katherine Ely, Traci Sitzmann, and Cari Falkiewicz, "The Influence of Goal Orientation Dimensions on Time to Train in a Self-Paced Training Environment," *Learning and Individual Differences*, Vol. 19, No. 1, 2009, pp. 146–150.

[222] Michael R. Kukenberger, John E. Mathieu, and Thomas Ruddy, "A Cross-Level Test of Empowerment and Process Influences on Members' Informal Learning and Team Commitment," *Journal of Management*, Vol. 41, No. 3, 2015, pp. 987–1016; Lee Di Milia and Kamal Birdi, "The Relationship Between Multiple Levels of Learning Practices and Objective and Subjective Organizational Financial Performance," *Journal of Organizational Behavior*, Vol. 31, No. 4, 2010, pp. 481–498.

and social learning.[223] Meta-analyses have provided evidence for a wide variety of training methods.[224] However, proper design, delivery, and implementation can enhance a given method's efficacy.[225]

The science of workplace training is the application of evidence-based principles that guide the structure and design of training. Workplace science aids researchers and practitioners in choosing features of training programs, including attendance requirements, trainee characteristics, the number of and spacing of sessions, and the instructor. It can also inform the selection of design features related to implementation, such as training methods, the use of feedback, the source of feedback, the training location, and the training setting (i.e., technology-based, in-person, or blended).[226] Kraiger and Ford,[227] drawing on several meta-analytic reviews, identify five empirically supported principles of workplace instruction to support the process of training design:

1. **Organize content in ways meaningful to learners.** Well-organized content contains information that is concise, coherent, and well connected. Including an instructional aid is also recommended.
2. **Optimize the sequencing of training so that content actively engages the learner.** This may involve scaffolding task complexity, increasing difficulty based on mastery of earlier phases, and implementing a practice schedule that mixes diverse content.
3. **Engage learners in their own learning experience.** Current supported practices for engaging learners include prompting learners to relate new knowledge to existing knowledge, asking learners to connect new concepts to prior knowledge, and having learners recall previously learned concepts.
4. **Provide effective practice for learners.** The literature recommends that practice is spaced over multiple sessions. It is also recommended that learners have opportunities to apply skills in diverse situations that mimic the performance environment.
5. **Develop past initial mastery.** Ensuring that learners understand the correct results, encouraging practice beyond the point where mastery has been achieved, and providing feedback are effective methods for learners to advance beyond initial mastery.

In the sections below, additional factors central to training design are described in further detail.

---

[223] Bradley L. Kirkman, John E. Mathieu, John L. Cordery, Benson Rosen, and Michael Kukenberger, "Managing a New Collaborative Entity in Business Organizations: Understanding Organizational Communities of Practice Effectiveness," *Journal of Applied Psychology*, Vol. 96, No. 6, 2011, p. 1234–1245.

[224] Salas, DiazGranados et al., 2008; Lisa A. Delise, C. Allen Gorman, Abby M. Brooks, Joan R. Rentsch, and Debra Steele-Johnson, "The Effects of Team Training on Team Outcomes: A Meta-Analysis," *Performance Improvement Quarterly*, Vol. 22, No. 4, 2010, pp. 53–80; Ashley M. Hughes, Megan E. Gregory, Dana L. Joseph, Shirley C. Sonesh, Shannon L. Marlow, Christina N. Lacerenza, Lauren E. Benishek, Heidi B. King, and Eduardo Salas, "Saving Lives: A Meta-Analysis of Team Training in Healthcare," *Journal of Applied Psychology*, Vol. 101, No. 9, 2016, pp. 1266–1304.

[225] Eduardo Salas, Scott I. Tannenbaum, Kurt Kraiger, and Kimberly A. Smith-Jentsch, "The Science of Training and Development in Organizations: What Matters in Practice," *Psychological Science in the Public Interest*, Vol. 13, No. 2, 2012, pp. 74–101.

[226] Lacerenza et al., 2017.

[227] Kraiger and Ford, 2021.

Trainee Characteristics

Trainee characteristics broadly encompass a trainee's personality traits, capabilities, attitudes and emotions, motivational constructs, values and interests, and perceptions.[228] Research has identified trainee characteristics that influence motivation to participate in training and apply learned skills afterward.[229] Although trainee characteristics are generally less amenable to change through training interventions, effective training design considers these characteristics to increase trainee motivation and enhance the likelihood of transferring skills beyond the training environment.

For example, social learning theory emphasizes that people learn by observing others whom they believe are credible. In this way, learning occurs by watching the consequences of another's behavior rather than having to directly experience the consequences themselves.[230] From this perspective, trainees may develop some amount of self-efficacy—the belief or judgment that they can successfully complete a task. Self-efficacy is an antecedent of learning, and it can be specific to a particular task.[231] That is, trainees can have high self-efficacy for performing one swim technique but not another. Instructional elements can be designed to increase self-efficacy; examples are verbal persuasion (offering encouragement),[232] logical verification (perceiving a relationship between a new task and a task already mastered),[233] modeling (peer demonstrations),[234] providing opportunities for practice,[235] gradually increasing task difficulty,[236] and ensuring that the training met the trainee's expectations.[237]

Goal-setting is another strategy that can be used to increase trainee motivation. To achieve high performance levels, individuals must be committed to stated goals, and the goals must be specific and

---

[228] Bell et al., 2017.

[229] Raymond A. Noe, "Trainees' Attributes and Attitudes: Neglected Influences on Training Effectiveness," *Academy of Management Review*, Vol. 11, No. 4, 1986, pp. 736–749; Jason A. Colquitt, Jeffrey A. LePine, and Raymond A. Noe, "Toward an Integrative Theory of Training Motivation: A Meta-Analytic Path Analysis of 20 Years of Research," *Journal of Applied Psychology*, Vol. 85, No. 5, 2000, pp. 678–707.

[230] Bandura and Walters, 1977.

[231] Albert Bandura, William H. Freeman, and Richard Lightsey, *Self-Efficacy: The Exercise of Control*, New York: Springer, 1999; Colquitt, LePine, and Noe, 2000; Lisa Dragoni, Haeseen Park, Jim Soltis, and Sheila Forte-Trammell, "Show and Tell: How Supervisors Facilitate Leader Development Among Transitioning Leaders," *Journal of Applied Psychology*, Vol. 99, No. 1, 2014, pp. 66–86.

[232] Albert Bandura, "The Explanatory and Predictive Scope of Self-Efficacy Theory," *Journal of Social and Clinical Psychology*, Vol. 4, No. 3, 1986, pp. 359–373.

[233] Bandura, 1986.

[234] Paul J. Taylor, Darlene F. Russ-Eft, and Daniel W. L. Chan, "A Meta-Analytic Review of Behavior Modeling Training," *Journal of Applied Psychology*, Vol. 90, No. 4, 2005, pp. 692–709.

[235] Gillian B. Yeo and Andrew Neal, "A Multilevel Analysis of Effort, Practice, and Performance: Effects; of Ability, Conscientiousness, and Goal Orientation," *Journal of Applied Psychology*, Vol. 89, No. 2, 2004, pp. 231–247.

[236] Michael G. Hughes, Eric Anthony Day, Xiaoqian Wang, Matthew J. Schuelke, Matthew L. Arsenault, Lauren N. Harkrider, and Olivia D. Cooper, "Learner-Controlled Practice Difficulty in the Training of a Complex Task: Cognitive and Motivational Mechanisms," *Journal of Applied Psychology*, Vol. 98, No. 1, 2013, pp. 80–98.

[237] Scott I. Tannenbaum, John E. Mathieu, Eduardo Salas, and Janis A. Cannon-Bowers, "Meeting Trainees' Expectations: The Influence of Training Fulfillment on the Development of Commitment, Self-Efficacy, and Motivation," *Journal of Applied Psychology*, Vol. 76, No. 6, 1991, pp. 759–769.

challenging but not perceived as unattainable.[238] If goals are perceived as too challenging, trainees may not be committed to pursuing those goals.[239] Hence, the implementation of clear action plans in training that set specific and challenging yet obtainable goals can contribute to successful outcomes.[240]

It is important to note that the influence of trainee characteristics on outcomes is nuanced. For example, although many studies have demonstrated the benefits of increasing self-efficacy among trainees, Vancouver and Kendall suggest that high self-efficacy may compel trainees to allocate fewer resources when planning, resulting in a negative relationship between self-efficacy and performance.[241] This finding suggests that the factors contributing to learning may vary among trainees. Therefore, training instructors should consider individual characteristics when structuring training and establishing goals.

Knowledge and Skill Acquisition

As the nature of work becomes increasingly complex, training objectives have shifted from optimizing the performance of simple manual tasks to learning how to complete multifaceted tasks and respond to dynamic, often ambiguous situations.[242] Although definitions vary,[243] competence is generally thought of as the integrated knowledge, attitudes, and skills needed to successfully perform a professional task.[244] Accordingly, the acquisition of knowledge and skill develops competence.

Skills refer to an organized sequence of activities that includes both the organization of movement (motor skills) and symbolic information (cognitive skills).[245] Often skills can be viewed as the application of knowledge. Effective knowledge and skill acquisition is a function of the match between the training method and the desired knowledge or skill.[246] For example, explicit knowledge, such as policies and procedures, can be transferred through traditional lecture-based instruction. In contrast,

---

[238] Edwin A. Locke and Gary P. Latham, "Goal Setting Theory," in *Motivation: Theory and Research*, New York: Routledge, 2012, pp. 23–40; Edwin A. Locke, Gary P. Latham, and Miriam Erez, "The Determinants of Goal Commitment," *Academy of Management Review*, Vol. 13, No. 1, 1988, pp. 23–39; Stephanie C. Payne, Satoris S. Youngcourt, and J. Matthew Beaubien, "A Meta-Analytic Examination of the Goal Orientation Nomological Net," *Journal of Applied Psychology*, Vol. 92, No. 1, 2007, pp. 128–150.

[239] Locke, Latham, and Erez, 1988.

[240] Eunju Choi and Douglas A. Johnson, "Common Antecedent Strategies Within Organizational Behavior Management: The Use of Goal Setting, Task Clarification, and Job Aids," *Journal of Organizational Behavior Management*, Vol. 42, No. 1, 2022, pp. 75–95; Fred C. Lunenburg, "Goal-Setting Theory of Motivation," *International Journal of Management, Business, and Administration*, Vol. 15, No. 1, 2011, pp. 1–6.

[241] Jeffrey B. Vancouver and Laura N. Kendall, "When Self-Efficacy Negatively Relates to Motivation and Performance in a Learning Context," *Journal of Applied Psychology*, Vol. 91, No. 5, 2006, pp. 1146–1153.

[242] Noe, Clarke, and Klein, 2014; Bell et al., 2017.

[243] Françoise Delamare Le Deist and Jonathan Winterton, "What Is Competence?" *Human Resource Development International*, Vol. 8, No. 1, 2005, pp. 27–46.

[244] Liesbeth K. J. Baartman and Elly De Bruijn, "Integrating Knowledge, Skills and Attitudes: Conceptualising Learning Processes Towards Vocational Competence," *Educational Research Review*, Vol. 6, No. 2, 2011, pp. 125–134.

[245] Michael I. Posner, Gregory J. DiGirolamo, and Diego Fernandez-Duque, "Brain Mechanisms of Cognitive Skills," *Consciousness and Cognition*, Vol. 6, No. 2–3, 1997, pp. 267–290.

[246] Arthur et al., 2003.

developing the tacit knowledge associated with when and how to apply, adopt, or abandon those practices may require alternative methods such as experiential learning.[247]

Whether retrieving a piece of information or executing a procedure, certain laws of skill and acquisition always apply,[248] including the law of practice and knowledge of results. The law of practice states that skill develops in a nonlinear way such that accuracy and speed improve quickly at first, but subsequent improvements are more gradual, requiring additional practice over time.[249] Several meta-analytic reviews support an association between practice-based delivery methods and desired training outcomes.[250] Practice-based delivery methods may include in-basket exercises, role-play, and simulations.

The conditions of practice are vital; practice without knowing if something was performed correctly can lead to students adopting incorrect strategies. This second major law of practice states that skill acquisition benefits from knowledge of results.[251] There is growing evidence that feedback accelerates the pace of learning. Explanatory feedback, which pushes past simply flagging errors to explain why something is incorrect, has been shown to enhance learning.[252]

Recommendations for when to provide feedback differs depending on the characteristics of the skill and the learner. Immediate feedback is necessary for certain contexts (i.e., to prevent an accident). Yet, immediate feedback reduces opportunities to develop self-regulated learning skills. Experiencing error allows trainees to anticipate potential problems and develop solutions to those problems.[253] Training design that includes multiple sessions is a valuable alternative to immediate feedback because it generates space for trainees to make errors. Instructors can offer delayed feedback in debriefs following sessions and support trainees in developing goals for future performance. In the following section, we review strategies of how to optimize feedback, practice, and other components of training design to increase the chances that training transfers to the job.

Transfer of Training

Learning does not always translate into behavior changes outside the classroom. To increase the likelihood that training transfers to the job, workplace science literature has focused on identifying key factors that influence the transfer of training. Figure A.1 illustrates the relationships between trainee

---

[247] Steven J. Armstrong and Anis Mahmud, "Experiential Learning and the Acquisition of Managerial Tacit Knowledge," *Academy of Management Learning & Education*, Vol. 7, No. 2, 2008, pp. 189–208.

[248] National Research Council, 2012.

[249] Paul Rosenbloom and Allen Newell, "Learning by Chunking: A Production System Model of Practice," in David Klahr, Patrick W. Langley, and Robert. T. Neches, eds., *Production System Models of Learning and Development*, Cambridge, Mass.: MIT Press, 1987, p. 221.

[250] Lacerenza et al., 2017; Salas, DiazGranados et al., 2008; Paul J. Taylor, Darlene F. Russ-Eft, and Hazel Taylor, "Transfer of Management Training from Alternative Perspectives," *Journal of Applied Psychology*, Vol. 94, No. 1, 2009, pp. 104–121; Michael J. Burke and Russell R. Day, "A Cumulative Study of the Effectiveness of Managerial Training," *Journal of Applied Psychology*, Vol. 71, No. 2, 1986, pp. 232–245.

[251] National Research Council, 2012.

[252] Roxana Moreno, "Decreasing Cognitive Load for Novice Students: Effects of Explanatory Versus Corrective Feedback in Discovery-Based Multimedia," *Instructional Science*, Vol. 32, No. 1, 2004, pp. 99–113.

[253] Nina Keith and Michael Frese, "Effectiveness of Error Management Training: A Meta-Analysis," *Journal of Applied Psychology*, Vol. 93, No. 1, 2001, pp. 59–69.

Figure A.1. Grossman and Salas Model of Training Transfer

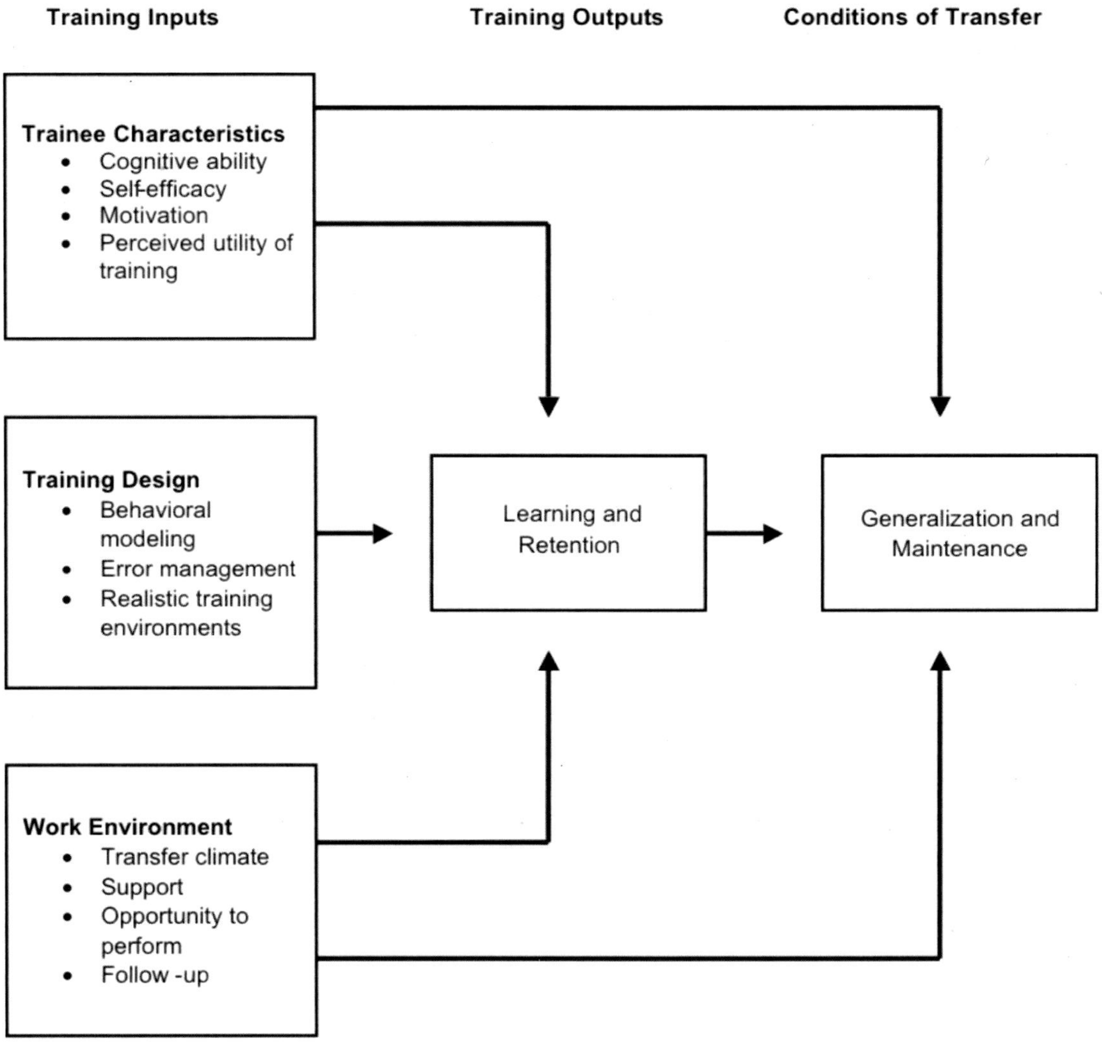

SOURCE: Rebecca Grossman and Eduardo Salas, "The Transfer of Training: What Really Matters," *International Journal of Training and Development*, Vol. 15, No. 2, 2011, pp. 103–120.

characteristics, training design, and the work environment on learning and retention and the transfer of training.[254] Many of these elements are captured in a concise checklist.[255] This checklist has guiding actions that organizations can take before, during, and after training:

Before Training

- Align training with objectives by conducting a training needs analysis.
- Prepare the environment for training by communicating the importance of training and establishing goals.

---

[254] Rebecca Grossman and Eduardo Salas, "The Transfer of Training: What Really Matters," *International Journal of Training and Development*, Vol. 15, No. 2, 2011, pp. 103–120.

[255] Ashley M. Hughes, Stephanie Zajac, Jacqueline M. Spencer, and Eduardo Salas, "A Checklist for Facilitating Training Transfer in Organizations," *International Journal of Training and Development*, Vol. 22, No. 4, 2018, pp. 334–345.

- Prepare and motivate trainees by communicating outcomes that can occur following training (i.e., benefits and consequences of performance).
- Develop a learning climate by allocating enough time and opportunity for trainees to learn, practice, and receive feedback.

During Training

- Provide multiple practice opportunities to encourage learning.
- Incorporate debriefs to provide constructive feedback on what actions trainees can adopt to improve their performance.
- Encourage trainees to seek feedback and to recognize that errors are learning opportunities.

After Training

- Establish a system to recognize and reward performance improvements.
- Monitor progress toward goals, and offer support through coaching/mentorship.
- Evaluate the effectiveness of training using multiple criteria determined by training needs analysis (e.g., learning and behavior change).
- Modify the training based on results from training evaluation and other lessons learned.

Appendix B

SWTW Document Review Methodology and Supplementary Analyses

Approach to Identifying Attributes in Course Documents

RAND's analysis of attributes found in SW training documents was conducted for the following courses: SWCC, PJ Apprentice, CCT Apprentice, SR Apprentice, TACP Apprentice, ATC, Pre-Dive, Open Circuit Dive, and Closed Circuit Dive. The five AFSC-specific courses are broken into training blocks—large portions of a course related to the same broad function or topic—whereas SWCC and the three diving-related courses each consist of a single block. Training blocks are further broken down into training units that provide additional details from which courses are structured. Finally, each training unit consists of one or more training objectives from which specific training events are designed.

These training objectives/events became the basis for much of RAND's analysis. It should be noted that each of the courses assessed has a different number of objectives/events specified in the training documentation, as represented in Figure B.1. These differences are mostly associated with the difference in overall course length but could also potentially occur if there are differences in how individual, entry-level tasks were defined and subsequently used as course objectives/events.

The AFSPECWAR training community provided RAND with over 500 training course documents. Some were technical in nature, such as specific training aids or evaluation processes, and others were more support-related logistics and safety measures, for example. Of all the documents reviewed, RAND relied primarily on the course charts, lesson plans, and POIs to understand each course's structure and content. These document types provided relevant information for analysis, including course structure, lesson content, and guidance for instructors in teaching and evaluating students during specific objectives/events. These details helped analysts better understand each course and develop an approach to analysis. For example, the POIs were particularly helpful in identifying what "training method" is applied to each objective or event. After loading a spreadsheet with a course's training blocks, units, and objectives/events as rows, the RAND team labeled each objective or event with one or more of five training methods as specified in the POIs:

- lecture/discussion (L/D)
- demonstration (Demo)
- performance–student practice (PR)
- performance evaluation (PE) or performance check (PC)
- exercise (EX).

Figure B.1. Number of Objectives/Events and Length of Courses Assessed

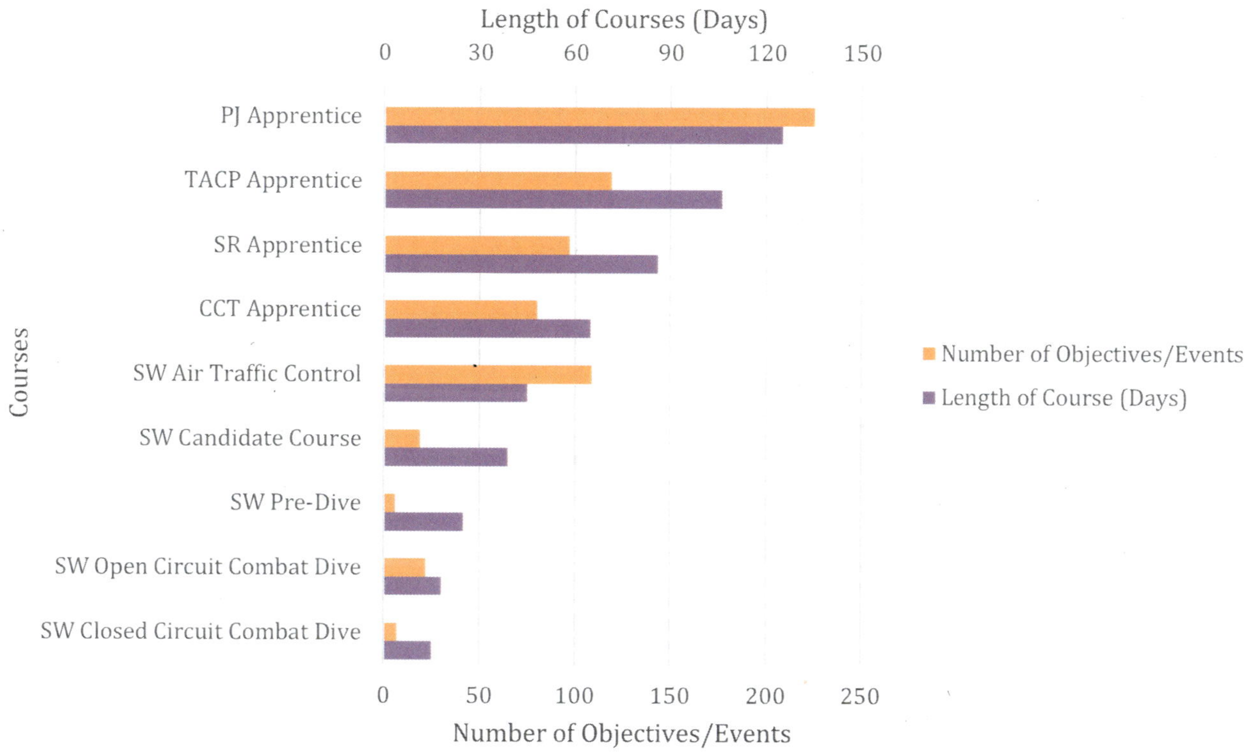

SOURCE: RAND analysis, using data from respective course documentation.

Figure B.2 portrays the proportion of the respective courses' objectives/events as represented by the five different training methods.

Figure B.2. Method of Instruction Across AFSPECWAR Initial Training Courses

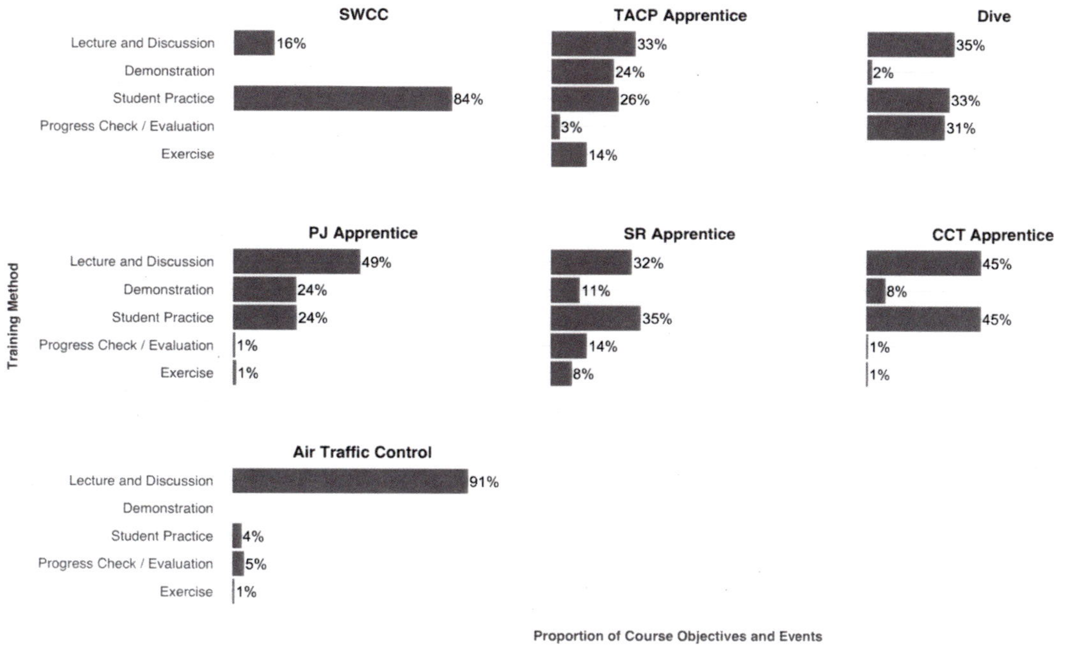

The 21 facets of AFSPECWAR's eight attributes (see Figure 1.1) were listed in columns for each of the course's respective spreadsheets. The RAND team then developed a set of numerical codes that could be used to assess whether a given facet was directly or indirectly developed, evaluated, had the potential for developing, or was unrelated to each objective or event. The numerical codes and their definitions are provided in Table B.1.

Table B.1. Numerical Codes and Definitions

Code #	Code Definition
0	Objective/event is assessed as not being directly/indirectly addressed (developed or evaluated) and unlikely an opportunity.
1	Facet is directly developed during objective/event.
2	Facet is directly evaluated during objective/event.
3	Facet is generally reinforced or indirectly developed during objective/event.
4	Opportunity exists for deliberate development or evaluation of the facet during objective/event.

Two RAND analysts then followed an inductive and iterative approach to coding with three consecutive methods:

1. Sort objectives/events by training method, and isolate those that had obvious connections to specific facets. For example, those objectives/events that only included a lecture/discussion could be coded as "directly developed" for the Learning Content facet of Trainability, investigated for other related facets, and then coded as a "0" for the remaining unrelated facets. A list of coding rules was developed for this approach but was not comprehensive enough to cover many combinations of objective, events, and facets.
2. Search through a course's objective and event titles using terms related to specific attributes. Those objectives/events with relevant search terms were then highlighted for the analyst to focus on for a particular attribute. For example, the terms "team," "small unit," and "leader" were quite useful for identifying objectives/events that should be closely considered for coding Teamwork facets. Some search terms were also combined with more prescriptive coding rules based on the combination of a search term and a specific training method. For example, in every case that "physical training" occurred in an event that was labeled PR for training method (student practice or practical application), every facet of Physical Fitness was coded as being directly developed. By inductively developing attribute-related search terms, analysts found that few common terms usefully reflected Drive or Integrity, though, and many objectives/events remained uncoded requiring a third approach.
3. After completing the previous two approaches, many objectives/events had still not been coded, so each one was reviewed systematically. Some coding rules were established for objectives/events that were common across courses. For example, those that included "live fire battle drills," "react to contact," and "patrol" could be coded similarly for the facets of Teamwork and Communication across courses. However, this method required the most in-depth reading of course materials and introduced the most subjectivity into the process. Overall, many of the coding rules facilitated finding strong connections between training events and the facets, although some were less prescriptive, leaving some discretion (e.g., direct

versus indirect development) to the coder to decide if attributes were developed. As described in Chapter 2, these difficulties were later reconciled by collapsing the indirect and direct development codes into one "developed" code.

The first two methods produced narrowly defined coding rules that analysts could apply independently. These methods also allowed for coders to discuss coding challenges and update rules as needed. Since the third approach left more room for coder subjectivity, the RAND team made additional efforts to reduce differences in how coders evaluated objectives/events. Both coders cross-checked courses they had not coded, discussing how each coder approached particularly challenging course material. After the initial round of coding was completed, graphical outputs for early analysis were used to identify outliers—facets with very high or low representation in a particular course, for example. Those were then further reviewed by both coders to determine if a coding error had occurred or if some aspect of the course explained the results. Updates to the original coding were required in some cases, although very rarely and mostly restricted to a few facets within one course. Overall, 14,532 codes were manually assigned, as displayed in Figure B.3.[256]

Figure B.3. Total Number of Objectives/Events Linked to Attribute Facets

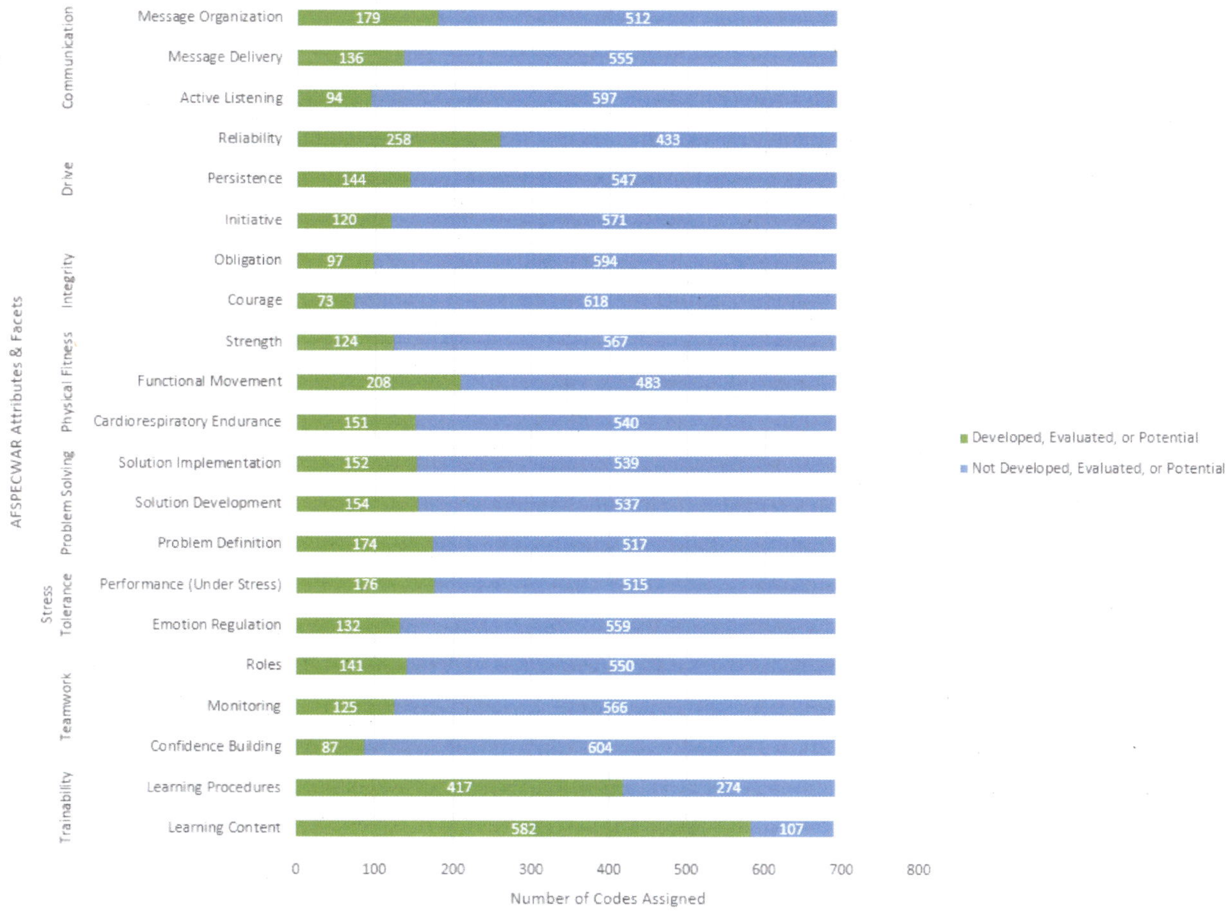

[256] Based on 692 objectives/events times 21 facets = 14,532 codes.

Appendix C

Attribute Literature Review Approach and Supplementary Findings

This report relies on peer-reviewed scholarly literature. We used a multistep search and filter process to identify literature on evidence-based practices for each AFSPECWAR attribute. For our initial search, we asked RAND Knowledge Services (KS) to search English-language sources from 1990–2021 using search terms in Table C.1 and the following search parameters:

- (training OR development OR malleability OR growth OR improvement)

AND

- ("meta-analysis" OR "meta analysis" OR "systematic review" OR "narrative review")

AND

- content terms.

Table C.1. Initial Attribute Search Terms

Attribute	Initial Search Terms
Teamwork	Influencing others; inspiring trust; interpersonal skills; interpersonal effectiveness; interpersonal adaptability; team building; conflict management; conflict resolution; consideration of others; awareness of others; social skills; teamwork
Stress Tolerance	Adaptability; resilience; situational flexibility; tolerance for uncertainty; stress management; coping with stress; risk taking; emotional stability; handling work stress; stress inoculation
Trainability	Trainability; metacognitive skills; feedback acceptance; feedback seeking; feedback orientation
Problem Solving	Judgment; decisionmaking; critical thinking; analytical thinking; creative thinking; innovative thinking; analysis; decisiveness; problem definition; problem generation; problem implementation
Communication	Oral presentation; presentation skills; oral defense; delivery; speech; active listening; nonverbal; speaking; listening
Physical Fitness	Physical ability; physical fitness; physical readiness; muscular strength; muscular endurance; cardiorespiratory endurance/fitness; functional movement; agility; flexibility
Integrity	Courage; professionalism; moral; ethics; ethical decisionmaking; moral judgment; honesty; trustworthiness
Drive	Perseverance; initiative; persistence; reliability; precision; motivation; intrinsic; tenacity; energy; results orientation; self-discipline; self-control; career commitment; grit; dedication; determination, dependability, responsibility, attention to detail; work ethic

NOTE: Databases searched included Academic Search Complete, Military Collection, Military Database, PsycINFO, PubMed, Scopus, Web of Science.

The initial search results returned many studies out of scope, particularly in terms of the populations included in the sources (e.g., elementary students). To narrow the search, we limited it to peer-reviewed research in PsycINFO (which includes journals most likely to publish relevant research to working populations, e.g., *Journal of Applied Psychology, Organizational Behavior and Human Decision Processes, Journal of the Academy of Management, Journal of Personality and Social Psychology, Psychological Bulletin, Personnel Psychology, Small Group Research,* and *Group Dynamics*) and Military Database.[257] Snowball (forward-sampling) approaches were also used to identify relevant sources cited in these articles that did not appear in the KS search results. Supplemental searches for meta-analyses and reviews in Google Scholar were also conducted to identify other relevant meta-analyses, reviews, and related studies that were not initially identified.

[257] A peer-reviewed delimiter is not available in DTIC.

Abbreviations

AFSPECWAR	Air Force Special Warfare
A&S	assessment and selection
ACC	Air Combat Command
AETC	Air Education and Training Command
AFSC	Air Force Specialty Code
AFSOC	Air Force Special Operations Command
AFSPECWAR	Air Force Special Warfare
AGOW	Air Ground Operations Wing
ATC	Air Traffic Control
BMT	Basic Military Training
CCT	Combat Control
CFETP	career field education and training plan
CMSgt	Chief Master Sargent
CRO	Combat Rescue Officer
CWB	counterproductive work behavior
DAF	Department of the Air Force
DSE	Developmental Special Experiences
EI	emotional intelligence
EX	exercise
FES	Feedback Environment Scale
FMP	full mission profile
FOS	Feedback Orientation Scale
FTU	formal training units
FY	fiscal year
GAWS	Guardian Angel Weapon System
HEXACO	Honesty–Humility, Emotionality, Extraversion, Agreeableness, Conscientiousness, and Openness to Experience
ICC	Initial Certification Course
IQC	Instructor Qualification Course
ISD	instructional system development
KSAOs	knowledge, skills, abilities, and other characteristics
L/D	lecture/discussion
MAJCOM	Major Command
NCO	non-commissioned officer

PAF	Project AIR FORCE
PC	performance check
PE	performance evaluation
PJ	PararescueJ
POIs	plans of instruction
PR	performance
PTSD	posttraumatic stress disorder
ROTC	Reserve Officer Training Corps
RQG	Rescue Group
SERE	Survival, Evasion, Resistance, and Escape
SES	Senior Executive Service
SIT	Stress Inoculation Training
SME	subject-matter expert
SOF	Special Operations Forces
SR	Special Reconnaissance
STO	Special Tactics Officer
STTS	Special Tactics Training Squadron
SW	Special Warfare
SWCC	Special Warfare Candidate Course
SWTW	Special Warfare Training Wing
TACP	Tactical Air Control Party
TDT	Team Dimensional Training
2AF	Second Air Force
UCMJ	Uniform Code of Military Justice
USAF	United States Air Force

References

Air Force Handbook (AFH) 33-337, *The Tongue and Quill*, Washington, D.C.: Secretary of the Air Force, May 27, 2015, Incorporating Change 1, November 19, 2015.

Allen, Joseph A., Roni Reiter-Palmon, John Crowe, and Cliff Scott, "Debriefs: Teams Learning from Doing in Context," *American Psychologist*, Vol. 73, No. 4, 2018, pp. 504–516.

Antony, Jesmin, Raman Brar, Paul A. Khan, Marco Ghassemi, Vera Nincic, Jane P. Sharpe, Sharon E. Straus, and Andrea C. Tricco, "Interventions for the Prevention and Management of Occupational Stress Injury in First Responders: A Rapid Overview of Reviews," *Systematic Reviews*, Vol. 9, No. 1, 2020, pp. 1–20.

Armstrong, Steven J., and Anis Mahmud, "Experiential Learning and the Acquisition of Managerial Tacit Knowledge," *Academy of Management Learning & Education*, Vol. 7, No. 2, 2008, pp. 189–208.

Arthur, Winfred, Jr., Winston Bennett, Jr., Pamela S. Edens, and Suzanne T. Bell, "Effectiveness of Training in Organizations: A Meta-Analysis of Design and Evaluation Features," *Journal of Applied Psychology*, Vol. 88, No. 2, 2003, pp. 234–245.

Ashman, Greg, Slava Kalyuga, and John Sweller, "Problem-Solving for Explicit Instruction: Which Should Go First When Element Interactivity Is High?" *Educational Psychology Review*, Vol. 32, 2020, pp. 229–247.

Ashton, Michael C. L., and Kibeom Lee, "Empirical, Theoretical, and Practical Advantages of the HEXACO Model of Personality Structure," *Personality and Social Psychology Review*, Vol. 11, No. 2, 2007, pp. 150–166.

Atwater, Leanne E., Joan F. Brett, and Atira Cherise Charles, "Multisource Feedback: Lessons Learned and Implications for Practice," *Human Resource Management*, Vol. 46, No. 2, 2007.

Atwater, Leanne E., Paul Roush, and Allison Fischthal, "The Influence of Upward Feedback on Self- and Follower Ratings of Leadership," *Personnel Psychology*, Vol. 48, No. 1, 1995.

Baartman, Liesbeth K. J., and Elly De Bruijn, "Integrating Knowledge, Skills and Attitudes: Conceptualising Learning Processes Towards Vocational Competence," *Educational Research Review*, Vol. 6, No. 2, 2011, pp. 125–134.

Bakken, Bjørn Tallak, and Martin Gilljam, "Dynamic Intuition in Military Command and Control: Why It Is Important, and How It Should Be Developed," *Cognition, Technology & Work*, Vol. 5, 2003, pp. 197–205.

Baldwin, Timothy T., and J. Kevin Ford, "Transfer of Training: A Review and Directions for Future Research," *Personnel Psychology*, Vol. 41, No. 1, 1988, pp. 63–105.

Bandura, Albert, "The Explanatory and Predictive Scope of Self-Efficacy Theory," *Journal of Social and Clinical Psychology*, Vol. 4, No. 3, 1986, pp. 359–373.

Bandura, Albert, "On the Functional Properties of Perceived Self-Efficacy Revisited," *Journal of Management*, Vol. 28, 2012, pp. 9–44.

Bandura, Albert, "Self-Efficacy: Toward a Unifying Theory of Behavioral Change," *Psychological Review*, Vol. 84, 1977, pp. 191–215.

Bandura, Albert, William H. Freeman, and Richard Lightsey, *Self-Efficacy: The Exercise of Control*, New York: Springer, 1999.

Bandura, Albert, and Edwin A. Locke, "Negative Self-Efficacy and Goal Effects Revisited," *Journal of Applied Psychology*, Vol. 88, 2003, pp. 87–99.

Bandura, Albert, and Richard H. Walters, *Social Learning Theory*, Vol. 1: Englewood Cliffs, N.J.: Prentice Hall, 1977.

Barrick, Murray R., and Michael K. Mount, "The Big Five Personality Dimensions and Job Performance: A Meta-Analysis," *Personnel Psychology*, Vol. 4, No. 1, 1991, pp. 1–26.

Barth, J., and P. Lannen, "Efficacy of Communication Skills Training Courses in Oncology: A Systematic Review and Meta-Analysis," *Annals of Oncology*, Vol. 22, 2011, pp. 1030–1040.

Baumann, Michael R., Carol L. Gohm, and Bryan L. Bonner, "Phased Training for High-Reliability Occupations: Live-Fire Exercises for Civilian Firefighters," *Human Factors*, Vol. 53, No. 5, 2011, pp. 548–557.

Bedi, Akanksha, Can M. Alpaslan, and Sandy Green, "A Meta-Analytic Review of Ethical Leadership Outcomes and Moderators," *Journal of Business Ethics*, Vol. 139, No. 3, 2016, pp. 517–536.

Bedwell, Wendy L., and Eduardo Salas, "Computer-Based Training: Capitalizing on Lessons Learned," *International Journal of Training and Development*, Vol. 14, No. 3, 2010, pp. 239–249.

Bell, Bradford S., and Steve W. J. Kozlowski, "Active Learning: Effects of Core Training Design Elements on Self-Regulatory Processes, Learning, and Adaptability," *Journal of Applied Psychology*, Vol. 93, No. 2, 2008, pp. 296–316.

Bell, Bradford S., Scott I. Tannenbaum, J. Kevin Ford, Raymond A. Noe, and Kurt Kraiger, "100 Years of Training and Development Research: What We Know and Where We Should Go," *Journal of Applied Psychology* (Centennial Special Issue), Vol. 102, No. 3, 2017, pp. 305–323.

Bell, James J., Lew Hardy, and Stuart Beattie, "Enhancing Mental Toughness and Performance Under Pressure in Elite Young Cricketers: A 2-Year Longitudinal Intervention," *Sport, Exercise, and Performance Psychology*, Vol. 2, 2013, pp. 281–297.

Berry, Christopher M., Deniz S. Ones, and Paul R. Sackett, "Interpersonal Deviance, Organizational Deviance, and Their Common Correlates: A Review and Meta-Analysis," *Journal of Applied Psychology*, Vol. 92, No. 2, 2007, pp. 410–424.

Beseler, Bradley, Christopher Mesagno, Warren Young, and Jack Harvey, "Igniting the Pressure Acclimatization Training Debate: Contradictory Pilot-Study Evidence from Australian Football," *Journal of Sport Behavior*, Vol. 39, 2016, pp. 22–38.

Birrer, Daniel, and Gareth Morgan, "Psychological Skills Training as a Way to Enhance an Athlete's Performance in High-Intensity Sports," *Scandinavian Journal of Medicine & Science in Sports*, Vol. 20, Supp. 2, 2010, pp. 78–87.

Bjornestad, Andrea, Seth Olson, and Lee Weidauer, "Improving Officer-Soldier Communications Through Active Listening Skills Training with Army ROTC Cadets," *Military Psychology*, Vol. 33, No. 3, 2021, pp. 205–215.

Blume, Brian D., J. Kevin Ford, Timothy T. Baldwin, and Jason L. Huang, "Transfer of Training: A Meta-Analytic Review," *Journal of Management*, Vol. 36, No. 4, 2010, pp. 1065–1105.

Bogg, Timothy, and Brent W. Roberts, "Conscientiousness and Health-Related Behaviors: A Meta-Analysis of the Leading Behavioral Contributors to Mortality," *Psychological Bulletin*, Vol. 130, No. 6, 2004, pp. 887–919.

Borenstein, Jason, Matthew J. Drake, Robert Kirkman, and Julie L. Swann, "The Engineering and Science Issues Test (ESIT): A Discipline-Specific Approach to Assessing Moral Judgment," *Science and Engineering Ethics*, Vol. 16, No. 2, 2010.

Bouchard, Stéphane, François Bernier, Éric Boivin, Brian Morin, and Geneviève Robillard, "Using Biofeedback While Immersed in a Stressful Videogame Increases the Effectiveness of Stress Management Skills in Soldiers," *PloS One*, Vol. 7, No. 4, 2012, e36169.

Brinkman, William B., Sheela R. Geraghty, Bruce P. Lanphear, Jane C. Khoury, Javier A. Gonzalez del Rey, Thomas G. DeWitt, Maria T. Britto, "Effect of Multisource Feedback on Resident Communication Skills and Professionalism: A Randomized Controlled Trial," *Archives of Pediatrics and Adolescent Medicine*, Vol. 161, January 2007.

Brown, Judith, "Training Needs Assessment: A Must for Developing an Effective Training Program," *Public Personnel Management*, Vol. 31, No. 4, 2002, pp. 569–578.

Brown, Michael E., Linda K. Treviño, and David A. Harrison, "Ethical Leadership: A Social Learning Perspective for Construct Development and Testing," *Organizational Behavior and Human Decision Processes*, Vol. 97, No. 2, 2005, pp. 117–134.

Burke, Michael J., and Russell R. Day, "A Cumulative Study of the Effectiveness of Managerial Training," *Journal of Applied Psychology*, Vol. 71, No. 2, 1986, pp. 232–245.

Burley, Simon D., Jace R. Drain, John A. Sampson, Bradley C. Nindl, and Herbert Groeller, "Effect of a Novel Low Volume, High Intensity Concurrent Training Regimen on Recruit Fitness and Resilience," *Journal of Science and Medicine in Sport*, Vol. 23, No. 10, pp. 2020979–2020984. As of February 15, 2024: https://doi.org/10.1016/j.jsams.2020.03.005

Cerasoli, Christopher P., "Antecedents and Outcomes of Informal Learning Behaviors: A Meta-Analysis," *Journal of Business and Psychology*, Vol. 33, 2018, pp. 203–230.

Cerasoli, Christopher P., Jessica M. Nicklin, and Michael T. Ford, "Intrinsic Motivation and Extrinsic Incentives Jointly Predict Performance: A 40-Year Meta-Analysis," *Psychological Bulletin*, Vol. 140, No. 4, 2014, pp. 980–1008.

Choi, Eunju, and Douglas A. Johnson, "Common Antecedent Strategies Within Organizational Behavior Management: The Use of Goal Setting, Task Clarification, and Job Aids," *Journal of Organizational Behavior Management*, Vol. 42, No. 1, 2022, pp. 75–95.

Colquitt, Jason A., Jeffrey A. LePine, and Raymond A. Noe, "Toward an Integrative Theory of Training Motivation: A Meta-Analytic Path Analysis of 20 Years of Research," *Journal of Applied Psychology*, Vol. 85, No. 5, 2000, pp. 678–707.

Connor, Kathryn M., and Jonathan R. T. Davidson, "Development of a New Resilience Scale: The Connor-Davidson Resilience Scale (CD-RISC)," *Depression and Anxiety*, Vol. 18, No. 2, 2003.

Costigan, Robert D., and Kyle E. Brink, "Developing Listening and Oral Expression Skills: Pillars of Influential Oral Communication," *Journal of Management Education*, Vol. 44, No. 2, 2020, pp. 129–164.

Crum, Alia, "Evaluating a Mindset Training Program to Unleash the Enhancing Nature of Stress," in *Academy of Management Proceedings*, Vol. 2011, No. 1, Briarcliff Manor, N.Y.: Academy of Management, 2011, pp. 1–6.

Crum, Alia J., Modupe Akinola, Ashley Martin, and Sean Fath, "The Role of Stress Mindset in Shaping Cognitive, Emotional, and Physiological Responses to Challenging and Threatening Stress," *Anxiety, Stress, & Coping*, Vol. 30, No. 4, 2017, pp. 379–395.

Dalal, Reeshad S., and Balca Bolunmez, "Evidence-Based Strategies to Improve Workplace Decisions: Small Steps, Big Effects," Alexandria, Va: Society for Human Resources Management, 2016. As of February 15, 2024:
https://www.siop.org/Portals/84/docs/SIOP-SHRM%20White%20Papers/SHRM-SIOP_Decision_Making.pdf

Davidson, Janet E., Rebecca Deuser, and Robert J. Sternberg, "The Role of Metacognition in Problem Solving," in J. Metcalfe and A. P. Shimamura, eds., *Metacognition: Knowing About Knowing*, MIT Press, 1994.

Delise, Lisa A., C. Allen Gorman, Abby M. Brooks, Joan R. Rentsch, and Debra Steele-Johnson, "The Effects of Team Training on Team Outcomes: A Meta-Analysis," *Performance Improvement Quarterly*, Vol. 22, No. 4, 2010, pp. 53–80.

Di Milia, Lee, and Kamal Birdi, "The Relationship Between Multiple Levels of Learning Practices and Objective and Subjective Organizational Financial Performance," *Journal of Organizational Behavior*, Vol. 31, No. 4, 2010, pp. 481–498.

Dilchert, Stephan, Deniz S. Ones, Robert D. Davis, and Cary D. Rostow, "Cognitive Ability Predicts Objectively Measured Counterproductive Work Behaviors," *Journal of Applied Psychology*, Vol. 92, No. 3, 2007, pp. 616–627.

Dixon, Raymond A., and Ryan A. Brown, "Transfer of Learning: Connecting Concepts During Problem Solving," *Journal of Technology Education*, Vol. 24, No. 1, Fall 2012, pp. 2–17.

Dragoni, Lisa, Haeseen Park, Jim Soltis, and Sheila Forte-Trammell, "Show and Tell: How Supervisors Facilitate Leader Development Among Transitioning Leaders," *Journal of Applied Psychology*, Vol. 99, No. 1, 2014, pp. 66–86.

Driskell, James E., and Eduardo Salas, "Collective Behavior and Team Performance," *Human Factors*, Vol. 34, No. 3, 1992.

Driskell, James E., et al., *Team Situational Awareness: A Research Model and Preliminary Analyses*, Orlando, Fla: Naval Air Warfare Training Systems Division, 1997.

Dweck, Carol S., *Mindset: The New Psychology of Success*, New York: Random House, 2006.

Edmonds, Grant W., Joshua J. Jackson, Jennifer V. Fayard, and Brent W. Roberts, "Is Character Fate, or Is There Hope to Change My Personality Yet?" *Social and Personality Psychology Compass*, Vol. 2, No. 1, 2008, pp. 399–413.

Ehrlich, Rachel J., Melissa R. Nosik, James E. Carr, and Byron Wine, "Teaching Employees How to Receive Feedback: A Preliminary Investigation," *Journal of Organizational Behavior Management*, Vol. 40, No. 1–2, 2020, pp. 19–29.

Ely, Katherine, Traci Sitzmann, and Cari Falkiewicz, "The Influence of Goal Orientation Dimensions on Time to Train in a Self-Paced Training Environment," *Learning and Individual Differences*, Vol. 19, No. 1, 2009, pp. 146–150.

Fishbach, Ayelet, and Kaitlin Woolley, "The Structure of Intrinsic Motivation," *Annual Review of Organizational Psychology and Organizational Behavior*, Vol. 9, 2022, pp. 339–363.

Flavell, John H., "Metacognition and Cognitive Monitoring: A New Area of Cognitive-Developmental Inquiry," *American Psychologist*, Vol. 34, No. 10, 1979.

Foote, Laura M., "Honing Crisis Communication Skills: Using Interactive Media and Student-Centered Learning to Develop Agile Leaders," *Journal of Management Education*, Vol. 37, No. 1, 2013, pp. 79–114.

Frayne, Colette A., and J. Michael Geringer, "Self-Management Training for Improving Job Performance: A Field Experiment Involving Salespeople," *Journal of Applied Psychology*, Vol. 85, No. 3, 2000, pp. 361–372.

Friborg, Oddgeir, Odin Hjemdal, Jan H. Rosenvinge, and Monica Martinussen, "A New Rating Scale for Adult Resilience: What Are the Central Protective Resources Behind Healthy Adjustment?" *International Journal of Methods in Psychiatric Research*, Vol. 12, No. 2, 2003.

Goldstein, Irwin L., "Training in Work Organizations," *Annual Review of Psychology*, Vol. 31, No. 1, 1980, pp. 229–272.

Gonzalez, Gabriella C., Reema Singh, Terry L. Schell, and Robin M. Weinick, *An Evaluation of the Implementation and Perceived Utility of the Airman Resilience Training Program*, Santa Monica, Calif.: RAND, RR-655-OSD, 2014. As of February 15, 2024:
https://www.rand.org/pubs/research_reports/RR655.html

Goodwin, Gerald F., Nikki Blacksmith, and Meredith R. Coats, "The Science of Teams in the Military: Contributions from over 60 Years of Research," *American Psychologist*, Vol. 73, No. 4, 2018, pp. 322–333.

Gore, Julie, Rhona Flin, Seville Stanton, and B. L. William Wong, "Applications for Naturalistic Decision-Making," *Journal of Occupational and Organizational Psychology*, Vol. 88, 2015, pp. 223–230.

Grgic, Jozo, Luke C. Mcllvenna, Jackson J. Fyfe, Filip Sabol, David J. Bishop, Brad J. Schoenfeld, and Zeljko Pedisic, "Does Aerobic Training Promote the Same Skeletal Muscle Hypertrophy as Resistance Training? A Systematic Review and Meta-Analysis," *Sports Medicine*, Vol. 49, No. 2, 2019, pp. 233–254.

Grossman, Rebecca, and Eduardo Salas, "The Transfer of Training: What Really Matters," *International Journal of Training and Development*, Vol. 15, No. 2, 2011, pp. 103–120.

Hallinger, Philip, and Edwin M. Bridges, "A Systematic Review of Research on the Use of Problem-Based Learning in the Preparation and Development of School Leaders," *Educational Administration Quarterly*, Vol. 53, No. 2, 2017, pp. 255–288.

Hattie, John, John Biggs, and Nola Purdie. "Effects of Learning Skills Interventions on Student Learning: A Meta-Analysis." *Review of Educational Research*, Vol. 66, No. 2, 1996, pp. 99–136.

Hatzigeorgiadis, Antonis, Nikos Zourbanos, Evangelos Galanis, and Yiannis Theodorakis, "Self-Talk and Sports Performance: A Meta-Analysis," *Perspectives on Psychological Science*, Vol. 6, No. 4, 2011, pp. 348–356.

Hourani, Laurel, Stephen Tueller, Paul Kizakevich, Gregory Lewis, Laura Strange, Belinda Weimer, Stephanie Bryant, Ellen Bishop, Robert Hubal, and James Spira, "Toward Preventing Post-Traumatic Stress Disorder: Development and Testing of a Pilot Predeployment Stress Inoculation Training Program," *Military Medicine*, Vol. 181, No. 9, 2016, pp. 1151–1160.

Hourani, Laurel, Stephen Tueller, Paul Kizakevich, Laura Strange, Gregory Lewis, Belinda Weimer, Jessica Morgan, Darryl Cooney, and Jessica Nelson, "Effect of Stress Inoculation Training with Relaxation Breathing on Perceived Stress and Posttraumatic Stress Disorder in the Military: A Longitudinal Study," *International Journal of Stress Management*, Vol. 25, No. S1, 2018, pp. 124–136.

HQ AETC A3/6. *Development of Air Force Foundational Competency Assessments Technical Report No. 2.*

Hu, Jinwei, Robert Lee, Sarah Mullin, Steven Schwaitzberg, Larry Harmon, Paul Gregory, and Peter L. Elkin, "How Physicians Change: Multisource Feedback Driven Intervention Improves Physician Leadership and Teamwork," *Surgery*, Vol. 168, 2020.

Hughes, Ashley M., Megan E. Gregory, Dana L. Joseph, Shirley C. Sonesh, Shannon L. Marlow, Christina N. Lacerenza, Lauren E. Benishek, Heidi B. King, and Eduardo Salas, "Saving Lives: A Meta-Analysis of Team Training in Healthcare," *Journal of Applied Psychology*, Vol. 101, No. 9, 2016, pp. 1266–1304.

Hughes, Ashley M., Stephanie Zajac, Jacqueline M. Spencer, and Eduardo Salas, "A Checklist for Facilitating Training Transfer in Organizations," *International Journal of Training and Development*, Vol. 22, No. 4, 2018, pp. 334–345.

Hughes, Michael G., Eric Anthony Day, Xiaoqian Wang, Matthew J. Schuelke, Matthew L. Arsenault, Lauren N. Harkrider, and Olivia D. Cooper, "Learner-Controlled Practice Difficulty in The Training of a Complex Task: Cognitive and Motivational Mechanisms," *Journal of Applied Psychology*, Vol. 98, No. 1, 2013, pp. 80–98.

Jackson, Joshua J., Tim Bogg, Kate E. Walton, Dustin Wood, Peter D. Harms, Jennifer Lodi-Smith, Grant W. Edmonds, and Brent W. Roberts, "Not All Conscientiousness Scales Change Alike: A Multimethod, Multisample Study of Age Differences in the Facets of Conscientiousness," *Journal of Personality and Social Psychology*, Vol. 96, No. 2, 2009, pp. 446–459.

Jerkiewicz, Carole L., "The Phantom Code of Ethics and Public Sector Reform," *Journal of Public Affairs*, Vol. 6, No. 3, May 2002.

Jones, Frank E., Yueh-Jyh Jong, Mark H. Jordan, Edwin T. Parks, Abigail Gray, Deanna A. Paulk, Lawrence G. Ruggiero, Stanley W. Shrader, Leneva J. Schroeder, and Rex Jordan, *Assessing the Communication Skills Curricula of Air Force Professional Military Education Programs*, research paper, ACSC/DR/078/96-04, April 1996.

Jones, Rebecca J., Stephen A. Woods, and Yves R. F. Guillaume, "The Effectiveness of Workplace Coaching: A Meta-Analysis of Learning and Performance Outcomes from Coaching," *Journal of Occupational and Organizational Psychology*, Vol. 89, No. 2, 2016, pp. 249–277.

Jozwiak, Jim, "Teaching Problem-Solving Skills to Adults," *MPAEA Journal of Adult Education*, Vol. 33, No. 1, Spring 2004, pp. 19–34.

Judge, Timothy A., Chad A. Higgins, Carl J. Thoresen, and Murray R. Barrick, "The Big Five Personality Traits, General Mental Ability, and Career Success Across the Life Span," *Personnel Psychology*, Vol. 52, No. 3, 1999, pp. 621–652.

Kalyuga, Slava, Alexander Renkl, and Fred Paas, "Facilitating Flexible Problem Solving: A Cognitive Load Perspective," *Educational Psychology Review*, Vol. 22, 2010, pp. 175–186.

Kalyuga, Slava, and John Sweller, "Cognitive Load and Expertise Reveal," in K. A. Ericsson, R. R. Hoffman, A. Kozbelt, and A. M. Williams, eds., *The Cambridge Handbook of Expertise and Expert Performance*, 2nd ed., Cambridge University Press, 2018, pp. 793–811.

Kanfer, Ruth, and Phillip L. Ackerman, "Aging, Adult Development, and Work Motivation," *Academy of Management Review*, Vol. 29, 2004, pp. 440–458.

Katsarov, Johannes, Roberto Andorno, André Krom, and Mariëtte van den Hoven, "Effective Strategies for Research Integrity Training—a Meta-analysis," *Educational Psychology Review*, 2021, pp. 1–21.

Katz, Ian M., Rachel S. Rauvola, and Cort W. Rudolph, "Feedback Environment: A Meta-Analysis," *International Journal of Selection and Assessment*, No. 29, Vol. 3–4, 2021, pp. 305–325.

Kayes, D. Christopher, David Stirling, and Tjai M. Nielson, "Building Organizational Integrity," *Business Horizons*, Vol. 50, 2007.

Keith, Nina, and Michael Frese, "Effectiveness of Error Management Training: A Meta-Analysis," *Journal of Applied Psychology*, Vol. 93, No. 1, 2001, pp. 59–69.

Kerr, Debra, Joan Ostaszkiewicz, Trisha Dunning, Peter Martin, "The Effectiveness of Training Interventions on Nurses' Communication Skills: A Systematic Review," *Nurse Education Today*, Vol. 89, 2020, pp. 1–13.

Kim, Ji Hyun, Barry Gerhart, and Meiyu Fang, "Do Financial Incentives Help or Harm Performance in Interesting Tasks?" *Journal of Applied Psychology*, Vol. 107, No. 1, 2022, pp. 153–167.

Kirkman, Bradley L., John E. Mathieu, John L. Cordery, Benson Rosen, and Michael Kukenberger, "Managing a New Collaborative Entity in Business Organizations: Understanding Organizational Communities of Practice Effectiveness," *Journal of Applied Psychology*, Vol. 96, No. 6, 2011, pp. 1234–1245.

Kirkpatrick, Donald L., *Evaluating Training Programs: The Four Levels*. San Francisco: Berrett-Koehler, 1994.

Klein, Cameron, Deborah DiazGranados, Eduardo Salas, Huy Le, C. Shawn Burke, Rebecca Lyons, and Gerald F. Goodwin, "Does Team Building Work?" *Small Group Research*, Vol. 40, No. 2, 2009, pp. 181–222.

Kluger, Avraham N. and Angelo DeNisi, "The Effects of Feedback Interventions on Performance: A Historical Review: A Meta-Analysis, and a Preliminary Feedback Intervention Theory," *Psychological Bulletin*, Vol. 119, No. 2, 1996, pp. 254–284.

Knapik, Joseph J., Everett A. Harman, Ryan A. Steelman, and Bria S. Graham, "A Systematic Review of the Effects of Physical Training on Load Carriage Performance," *Journal of Strength and Conditioning Research*, Vol. 26, No. 2, 2012, pp. 585–597. As of February 15, 2024: https://doi.org/10.1519/JSC.0b013e3182429853

Kolb, David A., *Experiential Learning: Experience as the Source of Learning and Development*, Upper Saddle River, N.J.: FT Press, 2014.

Kraiger, Kurt, Lisa M. Finkelstein, and Lebena S. Varghese, "Enacting Effective Mentoring Behaviors: Development and Initial Investigation of the Cuboid of Mentoring," *Journal of Business and Psychology*, Vol. 34, No. 4, 2019, pp. 403–424.

Kraiger, Kurt, and J. Kevin Ford, "The Science of Workplace Instruction: Learning and Development Applied to Work," *Annual Review of Organizational Psychology and Organizational Behavior*, Vol. 8, No. 1, 2021, pp. 45–72.

Krueger, Tracy C., Sean Robson, Joshua Snoke, Matthew Walsh, Matthew Strawn, Anthony Atler, Jason H. Campbell, Isabel Leamon, Ryan Haberman, and Barbara A. Bicksler, *Assessment and Selection for U.S. Air Force Special Warfare:* Vol. 1, *Defining Attributes and Designing Rater Training*, Santa Monica, Calif.: RAND, RR-A549-1, 2022. As of April 27, 2024: https://www.rand.org/pubs/research_reports/RRA549-1.html

Krueger, Tracy C., Sean Robson, Matthew Walsh, Anthony Atler, Beverly Weidmer, Matthew Strawn, Lewis Schneider, Lisa Pelled Colabella, Yuliya Shikh, and Tiffany Berglund, *Assessment and Selection for U.S. Air Force Special Warfare:* Vol. 2, *Evaluating Objectives and Considering Improvements*, Santa Monica, Calif.: RAND, RR-A1199-1, 2024. As of November 6, 2024: https://www.rand.org/pubs/research_reports/RRA1199-1.html

Kukenberger, Michael R., John E. Mathieu, and Thomas Ruddy, "A Cross-Level Test of Empowerment and Process Influences on Members' Informal Learning and Team Commitment," *Journal of Management*, Vol. 41, No. 3, 2015, pp. 987–1016.

Kyröläinen, Heikki, Kai Pihlainen, Jani P. Vaara, Tommi Ojanen, and Matti Santtila, "Optimising Training Adaptations and Performance in Military Environment," *Journal of Science and Medicine in Sport*, Vol. 21, No. 11 2018, pp. 1131–1138.

Lacerenza, Christina N., Shannon L. Marlow, Scott I. Tannenbaum, and Eduardo Salas, "Team Development Interventions: Evidence-Based Approaches for Improving Teamwork," *American Psychologist*, Vol. 73, No. 4, 2018, pp. 517–531.

Lacerenza, Christina N., Denise L. Reyes, Shannon L. Marlow, Dana L. Joseph, and Eduardo Salas, "Leadership Training Design, Delivery, and Implementation: A Meta-Analysis, *Journal of Applied Psychology*, Vol 102, No. 12, 2017.

Lachman, Vicki D., "Moral Courage: A Virtue in Need of Development?" *MESURG Nursing*, Vol. 16, No. 2, 2007, pp. 131–133.

Le Deist, Françoise Delamare, and Jonathan Winterton, "What Is Competence?" *Human Resource Development International*, Vol. 8, No. 1, 2005, pp. 27–46.

Lecours, Alexandra, Nancy Baril, and Marie-Josée Drolet, "What Is Professionalism in Occupational Therapy? A Concept Analysis: Qu'est-ce que le professionnalisme en ergothérapie? Analyse de ce concept," *Canadian Journal of Occupational Therapy*, Vol. 88, No. 2, 2021, pp. 117–130.

Lee, Kibeom, Michael C. Ashton, and Reinout E. de Vries, "Predicting Workplace Delinquency and Integrity with the HEXACO and Five-Factor Models of Personality Structure," *Human Performance*, Vol. 18, No. 2, 2005, 179–197.

Lesser, Cara S., Catherine R. Lucey, Barry Egener, Clarence H. Braddock, Stuart L. Linas, and Wendy Levinson, "A Behavioral and Systems View of Professionalism," *Journal of the American Medical Association*, Vol. 304, No. 24, 2010, pp. 2732–2737.

Lewis, Gregory F., Laurel Hourani, Stephen Tueller, Paul Kizakevich, Stephanie Bryant, Belinda Weimer, and Laura Strange, "Relaxation Training Assisted by Heart Rate Variability Biofeedback: Implication for a Military Predeployment Stress Inoculation Protocol," *Psychophysiology*, Vol. 52, No. 9, 2015, pp. 1167–1174.

Linderbaum, Beth A., and Paul E. Levy, "The Development and Validation of the Feedback Orientation Scale (FOS)," *Journal of Management*, Vol. 36, No. 6, 2010, pp. 1372–1405.

Liu, Yong, and Attila Pásztor, "Effects of Problem-Based Learning Instructional Intervention on Critical Thinking in Higher Education: A Meta-Analysis," *Thinking Skills and Creativity*, Vol. 45, Art. 101069, September 2022, pp. 1–21.

Locke, Edwin A., and Gary P. Latham, "Building a Practically Useful Theory of Goal Setting and Task Motivation: A 35-Year Odyssey," *American Psychologist*, Vol. 57, 2002, pp. 705–717.

Locke, Edwin A., and Gary P. Latham, "Goal Setting Theory," in *Motivation: Theory and Research*, New York: Routledge, 2012, pp. 23–40.

Locke, Edwin A., and Gary P. Latham, "New Directions in Goal-Setting Theory," *Current Directions in Psychological Science*, Vol. 15, No. 5, 2006, pp. 265–268.

Locke, Edwin A., and Gary P. Latham, *A Theory of Goal Setting and Task Performance*, Englewood Cliffs, N.J.: Prentice-Hall, 1990.

Locke, Edwin A., Gary P. Latham, and Miriam Erez, "The Determinants of Goal Commitment," *Academy of Management Review*, Vol. 13, No. 1, 1988, pp. 23–39.

London, M., and J.W. Smither, "Feedback Orientation, Feedback Culture, and the Longitudinal Performance Management Performance Management Process," *Human Resource Management Review*, Vol. 12, No. 1, 2002, pp. 81–100.

Low, William R., G. R. H. Sandercock, Paul Freeman, Marie E. Winter, Joanne Butt, and Ian Maynard, "Pressure Training for Performance Domains: A Meta-Analysis," *Sport, Exercise, and Performance Psychology*, Vol. 10, No. 1, 2021, pp. 149–163.

Lunenburg, Fred C., "Goal-Setting Theory of Motivation," *International Journal of Management, Business, and Administration*, Vol. 15, No. 1, 2011, pp. 1–6.

Lussier, James W., Karol G. Ross, and Bob Mayes, "Coaching Techniques for Adaptive Thinking," paper presented at the meeting of the Interservice/Industry Training, Simulation, and Education Conference, Orlando, Fla., December 2000.

Maglione, Margaret A., Christine Chen, Armenda Bialas, Aneesa Motala, Joan Chang, Goke Akinniranye, and Susanne Hempel, *Stress Control for Military, Law Enforcement, and First Responders: A Systematic Review*, Santa Monica, Calif.: RAND, RR-A119-3, 2021. As of August 25, 2022: https://www.rand.org/pubs/research_reports/RRA119-3.html

Malaby, Thomas M., "Beyond Play: A New Approach to Games," *Games and Culture*, Vol. 2, No. 2, 2007, pp. 95–113.

Marcus, Bernd, and Heinz Schuler, "Antecedents of Counterproductive Behavior at Work: A General Perspective," *Journal of Applied Psychology*, Vol. 89, No. 4, 2004, pp. 647–660.

Martin, Lynne, William O'Keefe, Lacey Schmidt, Immanuel Barshi, and Robert Mauro, "Houston, We Have a Problem Solving Model for Training," *Journal of Organizational Psychology*, Vol. 12, No. 1, 2012, pp. 57–68.

Mattingly, Victoria, and Kurt Kraiger, "Can Emotional Intelligence Be Trained? A Meta-Analytical Investigation," *Human Resource Management Review*, Vol. 29, No. 2, 2019, pp. 140–155.

Mayberry, Paul W., Charles A. Goldman, Kimberly Jackson, Eric Hastings, Hannah Acheson-Field, and Anthony Lawrence, *Making the Grade: Integration of Joint Professional Military Education and Talent Management in Developing Joint Officers*, Santa Monica, Calif.: RAND, RR-A473-1, 2021. As of August 25, 2022: https://www.rand.org/pubs/research_reports/RRA473-1.html

McEwan, Desmond, Geralyn R. Ruissen, Mark A. Eys, Bruno D. Zumbo, and Mark R. Beauchamp, "The Effectiveness of Teamwork Training on Teamwork Behaviors and Team Performance: A Systematic Review and Meta-Analysis of Controlled Interventions," *PloS One*, Vol. 12, No. 1, 2017, e0169604.

McInerney, Sarah A., Edward Waldrep, and Charles C. Benight, "Resilience Enhancing Programs in the U.S. Military: An Exploration of Theory and Applied Practice," *Military Psychology*, 2022, pp. 1–12.

Medeiros, Kelsey E., Logan L. Watts, and Michael D. Mumford, "Thinking Inside the Box: Educating Leaders to Manage Constraints," in Chunfang Zhou, ed., *Handbook of Research on Creative Problem-Solving Skill Development in Higher Education*, Hershey, Pa.: IGI Global, 2017, pp. 25–50.

Meredith, Lisa S., Cathy D. Sherbourne, Sarah J. Gaillot, Lydia Hansell, Hans V. Ritschard, Andrew M. Parker, and Glenda Wrenn, *Promoting Psychological Resilience in the U.S. Military*, Santa Monica, Calif.: RAND, MG-996-OSD, 2011. As of February 15, 2024:
https://www.rand.org/pubs/monographs/MG996.html

Militello, Laura G., Christen E. Sushereba, Matthieu Branlat, Robert Bean, and Victor Finomore, "Designing for Military Pararescue: Naturalistic Decision-Making Perspective, Methods, and Frameworks," *Journal of Occupational and Organizational Psychology*, Vol. 88, 2015, pp. 251–272.

Mitchell, Terence R., and Denise Daniels, "Motivation," in Walter C. Borman, Daniel R. Ilgen, and Richard J. Klimoski, eds., *Handbook of Psychology*, Vol. 12: *Industrial and Organizational Psychology*, Hoboken, N.J.: John Wiley & Sons, 2003, pp. 225–254.

Moreau, David, "How Malleable Are Cognitive Abilities? A Critical Perspective on Popular Brief Interventions," *American Psychologist*, Vol. 77, No. 3, 2022, pp. 409–423.

Moreno, Roxana, "Decreasing Cognitive Load for Novice Students: Effects of Explanatory Versus Corrective Feedback in Discovery-Based Multimedia," *Instructional Science*, Vol. 32, No. 1, 2004, pp. 99–113.

Morin, Lucie, and Gary Latham, "The Effect of Mental Practice and Goal Setting as a Transfer of Training Intervention on Supervisors' Self-Efficacy and Communication Skills: An Exploratory Study," *Applied Psychology*, Vol. 49, No. 3, 2000, pp. 566–578.

Morreaele, Sherwyn P., David W. Worley, and Barbara Hugenberg, "The Basic Communication Course at Two- and Four-Year U.S. Colleges and Universities: Study VIII—The 40th Anniversary," *Communication Education*, Vol. 59, No. 4, 2010, pp. 405–430.

National Research Council, *Education for Life and Work: Developing Transferable Knowledge and Skills in the 21st Century*, Washington, D.C.: National Academies Press, 2012.

National Research Council (US) Committee on the Assessment of 21st Century Skills, "Assessing Intrapersonal Skills," in *Assessing 21st Century Skills: Summary of a Workshop*, Washington, D.C.: National Academies Press, 2011. As of September 28, 2022:
https://www.ncbi.nlm.nih.gov/books/NBK84217/

Nelson, Johnathan K., Stephen J. Zaccaro, and Jeffrey L. Herman. "Strategic Information Provision and Experiential Variety as Tools for Developing Adaptive Leadership Skills," *Consulting Psychology Journal: Practice and Research*, Vol. 62, No. 2, 2010.

Noe, Raymond A., "Trainees' Attributes and Attitudes: Neglected Influences on Training Effectiveness," *Academy of Management Review*, Vol. 11, No. 4, 1986, pp. 736–749.

Noe, Raymond A., Alena D. M. Clarke, and Howard J. Klein, "Learning in the Twenty-First-Century Workplace," *Annual Review of Organizational Psychology and Organizational Behavior*, Vol. 1, No. 1, 2014, pp. 245–275.

Orr, Robin M., Rodney Pope, Venerina Johnston, and Julia Coyle, "Load Carriage: Minimising Soldier Injuries Through Physical Conditioning-A Narrative Review," *Journal of Military and Veterans Health*, Vol. 18, No. 3 2010, pp. 31–38.

Orvis, Karin A., and Krista Langkamer Ratwani, "Leader Self-Development: A Contemporary Context for Leader Development Evaluation," *The Leadership Quarterly*, Vol. 21, No. 4, 2010, pp. 657–674.

O'Shea, Patrick G., James E. Driskell, Gerald F. Goodwin, Michelle L. Zbylut, and Stephanie M. Weiss, *Development of a Conditional Reasoning Measure of Team Orientation*, Washington, D.C.: American Institutes for Research, 2004.

Oudejans, Raôul R. D., "Reality-Based Practice Under Pressure Improves Handgun Shooting Performance of Police Officers," *Ergonomics*, Vol. 51, 2008, pp. 261–273.

Pallavicini, Federica, Luca Argenton, Nicola Toniazzi, Luciana Aceti, and Fabrizia Mantovani, "Virtual Reality Applications for Stress Management Training in the Military," *Aerospace Medicine and Human Performance*, Vol. 87, No. 12, 2016, pp. 1021–1030.

Patterson, George T., Irene W. Chung, and Philip W. Swan, "Stress Management Interventions for Police Officers and Recruits: A Meta-Analysis," *Journal of Experimental Criminology*, Vol. 10, No. 4, 2014, pp. 487–513.

Payne, Stephanie C., Satoris S. Youngcourt, and J. Matthew Beaubien, "A Meta-Analytic Examination of the Goal Orientation Nomological Net," *Journal of Applied Psychology*, Vol. 92, No. 1, 2007, pp. 128–150.

Pichler, Shaun, Gerard Beenen, and Stephen Wood, "Feedback Frequency and Appraisal Reactions: A Meta-Analytic Test of Moderators," *International Journal of Human Resource Management*, Vol. 31, No. 17, 2020, pp. 2238–2263.

Poropat, Arthur E., "A Meta-Analysis of the Five-Factor Model of Personality and Academic Performance," *Psychological Bulletin*, Vol. 135, No. 2, 2009, pp. 322–338.

Posner, Michael I., Gregory J. DiGirolamo, and Diego Fernandez-Duque, "Brain Mechanisms of Cognitive Skills," *Consciousness and Cognition*, Vol. 6, No. 2–3, 1997, pp. 267–290.

Qin, Yongjie, Yungui Wang, and Robert E. Floden, "The Effect of Problem-Based Learning on Improvement of the Medical Educational Environment: A Systematic Review and Meta-Analysis," *Medical Principles and Practice*, Vol. 25, August 2016, pp. 525–532.

Renshaw, Ian, Keith Davids, Duarte Araújo, Ana Lucas, William M. Roberts, Daniel J. Newcombe, and Benjamin Franks, "Evaluating Weaknesses of 'Perceptual-Cognitive Training' and 'Brain Training' Methods in Sport: An Ecological Dynamics Critique," *Frontiers in Psychology*, Vol. 9, Art. 2468, January 2019, pp. 1–14.

Roberts, Brent W., Oleksandr S. Chernyshenko, Stephen Stark, and Lewis R. Goldberg, "The Structure of Conscientiousness: An Empirical Investigation Based on Seven Major Personality Questionnaires," *Personnel Psychology*, Vol. 58, No. 1, 2005, pp. 103–139.

Roberts, Brent W., Kate E. Walton, and Wolfgang Viechtbauer, "Patterns of Mean-Level Change in Personality Traits Across the Life Course: A Meta-Analysis of Longitudinal Studies," *Psychological Bulletin*, Vol. 132, No. 1, 2006.

Robson, Sean, *Psychological Fitness and Resilience: A Review of Relevant Constructs, Measures, and Links to Well-Being*, Santa Monica, Calif.: RAND, RR-102-AF, 2014. As of August 30, 2022: https://www.rand.org/pubs/research_reports/RR102.html

Robson, Sean, and Thomas Manacapilli, *Enhancing Performance Under Stress: Stress Inoculation Training for Battlefield Airmen*, Santa Monica, Calif.: RAND, RR-750-AF, 2014. As of June 22, 2022: https://www.rand.org/pubs/research_reports/RR750.html

Rosenbloom, Paul, and Allen Newell, "Learning by Chunking: A Production System Model of Practice," in David Klahr, Patrick W. Langley, and Robert T. Neches, eds., *Production System Models of Learning and Development*, Cambridge, Mass.: MIT Press, 1987, p. 221.

Russell, Teresa L., Taylor E. Sparks, John P. Campbell, Kristina Handy, Peter Ramsberger, and James A. Grand, "Situating Ethical Behavior in the Nomological Network of Job Performance," *Journal of Business and Psychology*, Vol. 32, No. 3, 2017, pp. 253–271.

Salas, Eduardo, and Janis A. Cannon-Bowers, "The Science of Training: A Decade," *Annual Review of Psychology*, Vol. 52, 2001, pp. 471–499.

Salas, Eduardo, Deborah DiazGranados, Cameron Klein, C. Shawn Burke, Kevin C. Stagl, Gerald F. Goodwin, and Stanley M. Halpin, "Does Team Training Improve Team Performance? A Meta-Analysis," *Human Factors*, Vol. 50, No. 6, 2008, pp. 903–933.

Salas, Eduardo, Michael A. Rosen, C. Shawn Burke, and Gerald F. Goodwin, "The Wisdom of Collectives in Organizations: An Update of the Teamwork Competencies," in Eduardo Salas, Gerald F. Goodwin, and C. Shawn Burke, eds., *Team Effectiveness in Complex Organizations: Cross-Disciplinary Perspectives and Approaches*, New York: Routledge, 2008, pp. 73–114.

Salas, Eduardo, Scott I. Tannenbaum, Kurt Kraiger, and Kimberly A. Smith-Jentsch, "The Science of Training and Development in Organizations: What Matters in Practice," *Psychological Science in the Public Interest*, Vol. 13, No. 2, 2012, pp. 74–101.

Savery, John R., "Overview of Problem-Based Learning: Definitions and Distinctions," *Interdisciplinary Journal of Problem-Based Learning*, Vol. 1, No. 1, Spring 2006, pp. 9–20.

Schleicher, Deidra J., Heidi M. Baumann, David W. Sullivan, and Junhyok Yim, "Evaluating the Effectiveness of Performance Management: A 30-Year Integrative Conceptual Review," *Journal of Applied Psychology*, Vol. 104, No. 7, 2019, pp. 851–887.

Schumann, Moritz, Joshua F. Feuerbacher, Marvin Sünkeler, Nils Freitag, Bent R. Rønnestad, Kenji Doma, and Tommy R. Lundberg, "Compatibility of Concurrent Aerobic and Strength Training for Skeletal Muscle Size and Function: An Updated Systematic Review and Meta-Analysis," *Sports Medicine*, November, Vol. 10, 2021, pp. 1–12.

Shadrick, Scott B., James W. Lussier, and Christopher Fultz, *Accelerating the Development of Adaptive Performance: Validating the Think Like a Commander Training*, Arlington, Va.: U.S. Army Research Institute for the Behavioral and Social Sciences, ARI Research Report 1868, 2007.

Shadrick, Scott B., Brian T. Crabb, James W. Lussier, and Thomas J. Burke, *Positive Transfer of Adaptive Battlefield Thinking Skills*, Arlington, Va.: U.S. Army Research Institute for the Behavioral and Social Sciences, ARI Research Report 1873, 2007. As of February 15, 2024: https://apps.dtic.mil/sti/citations/ADA470215

Simonsmeier, Bianca A., Melina Andronie, Susanne Buecker, and Cornelia Frank, "The Effects of Imagery Interventions in Sports: A Meta-Analysis," *International Review of Sport and Exercise Psychology*, Vol. 14, No. 1, 2021, pp. 186–207.

Sinha, Tanmay, and Manu Kapur, "When Problem Solving Followed by Instruction Works: Evidence for Productive Failure," *Review of Educational Research*, Vol. 91, No. 5, October 2021, pp. 761–798.

Sitzmann, Traci, "A Meta-Analytic Examination of the Instructional Effectiveness of Computer-Based Simulation Games," *Personnel Psychology*, Vol. 64, No. 2, 2011, pp. 489–528.

Smith, Bruce W., Jeanne Dalen, Kathryn Wiggins, Erin Tooley, Paulette Christopher, and Jennifer Bernard, "The Brief Resilience Scale: Assessing the Ability to Bounce Back," *International Journal of Behavioral Medicine*, Vol. 15, No. 3, 2008.

Smith, Chelsea, Kenji Doma, Brian Heilbronn, and Anthony Leicht, "Effect of Exercise Training Programs on Physical Fitness Domains in Military Personnel: A Systematic Review and Meta-Analysis," *Military Medicine*, No. 9/10, September 2022, pp. 1065–1073.

Smith, Eric N., Michael D. Young, and Alia J. Crum, "Stress, Mindsets, and Success in Navy SEALs Special Warfare Training," *Frontiers in Psychology*, Vol. 10, 2020.

Smither, James W., Manuel London, and Richard R. Reilly, "Does Performance Improve Following Multisource Feedback? A Theoretical Model, Meta-Analysis, and Review of Empirical Findings." *Personnel Psychology*, Vol. 58, No. 1, 2005, pp. 33–66.

Smith-Jentsch, Kimberly A., Janis A. Cannon-Bowers, Scott I. Tannenbaum, and Eduardo Salas, "Guided Team Self-Correction: Impacts on Team Mental Models, Processes, and Effectiveness," *Small Group Research*, Vol. 39, No. 3, 2008, pp. 303–327.

Spector, Paul E., "The Relationship of Personality to Counterproductive Work Behavior (CWB): An Integration of Perspectives," *Human Resource Management Review*, Vol. 21, No. 4, 2011, pp. 342–352.

Spiering, Barry A., Iñigo Mujika, Marilyn A. Sharp, and Stephen A. Foulis, "Maintaining Physical Performance: The Minimal Dose of Exercise Needed to Preserve Endurance and Strength over Time," *Journal of Strength and Conditioning Research*, Vol. 35, No. 5, 2021, pp. 1449–1558.

Steelman, Lisa A., Paul E. Levy, and Andrea F. Snell, "The Feedback Environment Scale: Construct Definition, Measurement, and Validation," *Educational and Psychological Measurement*, Vol. 64, No. 1, 2004, pp. 165–184.

Straus, Susan G., Michael G. Shanley, Carra S. Sims, Bryan W. Hallmark, Anna R. Saavedra, Stoney Trent, and Sean Duggan, *Innovative Leader Development: Evaluation of the U.S. Army Asymmetric Warfare Adaptive Leader Program*, Santa Monica, Calif.: RAND Arroyo Center, RR-504-A, 2014. As of February 14, 2024: https://www.rand.org/pubs/research_reports/RR504.html

Sweller, John, "Cognitive Load During Problem Solving: Effects on Learning," *Cognitive Science*, Vol. 12, 1988, pp. 257–285.

Sweller, John, Jeroen J. G. van Merriënboer, and Fred Paas, "Cognitive Architecture and Instructional Design: 20 Years Later," *Educational Psychology Review*, Vol. 31, 2019, pp. 261–292.

Tannenbaum, Scott I., John E. Mathieu, Eduardo Salas, and Janis A. Cannon-Bowers, "Meeting Trainees' Expectations: The Influence of Training Fulfillment on the Development of Commitment, Self-Efficacy, and Motivation," *Journal of Applied Psychology*, Vol. 76, No. 6, 1991, pp. 759–769.

Taylor, Gabrielle, and Raimond Gaita, "Integrity," *Aristotelian Society Supplementary Volume*, Vol. 55, No. 1, 1981.

Taylor, Paul J., and Michael P. O'Driscoll, "A New Integrated Framework for Training Needs Analysis," *Human Resource Management Journal*, Vol. 8, No. 2, 1998, pp. 1–29.

Taylor, Paul J., Darlene F. Russ-Eft, and Daniel W. L. Chan, "A Meta-Analytic Review of Behavior Modeling Training," *Journal of Applied Psychology*, Vol. 90, No. 4, 2005, pp. 692–709.

Taylor, Paul J., Darlene F. Russ-Eft, and Hazel Taylor, "Transfer of Management Training from Alternative Perspectives," *Journal of Applied Psychology*, Vol. 94, No. 1, 2009, pp. 104–121.

Teding van Berkhout, Emily, and John M. Malouff, "The Efficacy of Empathy Training: A Meta-Analysis of Randomized Controlled Trials," *Journal of Counseling Psychology*, Vol. 63, No. 1, 2016, pp. 32–41.

Tiruneh, Dawit T., An Verburgh, and Jan Elen, "Effectiveness of Critical Thinking Instruction in Higher Education: A Systematic Review of Intervention Studies," *Higher Education Studies*, Vol. 4, No. 1, 2014, pp. 1–17.

Torrence, Brett S., Logan L. Watts, Tyler J. Mulhearn, Megan R. Turner, E. Michelle Todd, Michael D. Mumford, and Shane Connelly, "Curricular Approaches in Research Ethics Education: Reflecting on More and Less Effective Practices in Instructional Content," *Accountability in Research*, Vol. 24, No. 5, 2017.

Toth, Adam J., Eoghan McNeill, Kevin Hayes, Aidan P. Moran, and Mark Campbell, "Does Mental Practice Still Enhance Performance? A 24 Year Follow-Up and Meta-Analytic Replication and Extension," *Psychology of Sport and Exercise*, Vol. 48, 2020, pp. 1–13.

Treviño, Linda Klebe, Laura Pincus Hartman, and Michael Brown, "Moral Person and Moral Manager: How Executives Develop a Reputation for Ethical Leadership," *California Management Review*, Vol. 42, No. 4, 2000, pp. 128–142.

U.S. Air Force, Airman Comprehensive Assessment Addendum, Air Force Form 724-A.

U.S. Air Force, Basic Military and Technical Training, Air Education and Training Command Instruction 36-2651, April 12, 2021.

U.S. Air Force, Competency Modeling, Air Force Handbook 36-2647, February 8, 2022.

U.S. Air Force, Information for Designers of Instructional Systems, Air Force Handbook 36-2235, November 1, 2002.

Vaara, Jani P., Herbert Groeller, Jace Drain, Heikki Kyröläinen, Kai Pihlainen, Tommi Ojanen, Chris Connaboy, Matti Santtila, Philip Agostinelli, and Brad C. Nindl, "Physical Training Considerations for Optimizing Performance in Essential Military Tasks," *European Journal of Sport Science*, Vol. 22, No. 1, 2022, pp. 43–57.

Vancouver, Jeffrey B., and David V. Day, "Industrial and Organisation Research on Self-Regulation: From Constructs to Application," *Applied Psychology: An International Review*, Vol. 54, No. 2, 2005, pp. 155–185.

Vancouver, Jeffrey B., and Laura N. Kendall, "When Self-Efficacy Negatively Relates to Motivation and Performance in a Learning Context," *Journal of Applied Psychology*, Vol. 91, No. 5, 2006, pp. 1146–1153.

Van den Broeck, Anja, Joshua L. Howard, Yves Van Vaerenbergh, Hannes Leroy, and Marylene Gagne, "Beyond Intrinsic and Extrinsic Motivation: A Meta-Analysis on Self-Determination Theory's Multidimensional Conceptualization of Work Motivation," *Organizational Psychology Review*, Vol. 11, No. 3, 2021, pp. 240–273.

Van Iddekinge, Chad H., John D. Arnold, Herman Aguinis, Jonas W. B. Lang, and Filip Lievens, "Work Effort: A Conceptual and Meta-Analytic Review," *Journal of Management*, Vol. 49, No. 1, 2023, pp. 125–157. As of February 15, 2024: https://doi.org/10.1177/01492063221087641

Waples, Ethan P., Alison L. Antes, Stephen T. Murphy, Shane Connelly, and Michael D. Mumford, "A Meta-Analytic Investigation of Business Ethics Instruction," *Journal of Business Ethics*, Vol. 87, No. 1, 2009, pp. 133–151.

Ward, Paul, Damian Farrow, Kevin R. Harris, A. Mark Williams, David W. Eccles, and K. Anders Ericsson, "Training Perceptual-Cognitive Skills: Can Sport Psychology Research Inform Military Decision Training?" *Military Psychology*, Vol. 20, Suppl. 1, 2008, pp. S71–S102.

Warner, Christopher H., George N. Appenzeller, Angela Mobbs, Jessica R. Parker, and Carolynn M. Warner, "Effectiveness of Battlefield-Ethics Training During Combat Deployment: A Programme Assessment," *Lancet*, Vol. 378, 2011.

Watts, Logan L., Kelsey E. Medeiros, Tyler J. Mulhearn, Logan M. Steele, Shane Connelly, and Michael D. Mumford, "Are Ethics Training Programs Improving? A Meta-Analytic Review of Past and Present Ethics Instruction in the Sciences," *Ethics & Behavior*, Vol. 27, No. 5, 2017, pp. 351–384.

Wiernik, Brenton M., and Deniz S. Ones, "Ethical Employee Behaviors in the Consensus Taxonomy of Counterproductive Work Behaviors," *International Journal of Selection and Assessment*, Vol. 26, No. 1, 2018, pp. 36–48.

Wilkinson, Tim J., Winnie B. Wade, and L. Doug Knock, "A Blueprint to Assess Professionalism: Results of a Systematic Review," *Academic Medicine*, Vol. 84, No. 5, 2009, pp. 551–558.

Wood, Robert, and Albert Bandura, "Impact of Conceptions of Ability on Self-Regulatory Mechanisms and Complex Decision Making," *Journal of Personality and Social Psychology*, Vol. 56, No. 3, 1989, pp. 407–415.

Yeo, Gillian B., and Andrew Neal, "A Multilevel Analysis of Effort, Practice, and Performance: Effects; of Ability, Conscientiousness, and Goal Orientation," *Journal of Applied Psychology*, Vol. 89, No. 2, 2004, pp. 231–247.

Zack, Michael H., "Managing Codified Knowledge," *Sloan Management Review*, Vol. 40, No. 4, 1999, pp. 45–58.